Generation of Polymers and Nanomaterials at Liquid–Liquid Interfaces

Generation of Polymers and Nanomaterials at Liquid–Liquid Interfaces

Application to Crystalline, Light Emitting, and Energy Materials

Second Edition

PANAGIOTIS DALLAS

Institute of Nanoscience and Nanotechnology,
NCSR Demokritos, Athens, Greece

ELSEVIER

Elsevier
Radarweg 29, PO Box 211, 1000 AE Amsterdam, Netherlands
The Boulevard, Langford Lane, Kidlington, Oxford OX5 1GB, United Kingdom
50 Hampshire Street, 5th Floor, Cambridge, MA 02139, United States

British Library Cataloguing-in-Publication Data
A catalogue record for this book is available from the British Library

Library of Congress Cataloging-in-Publication Data
A catalog record for this book is available from the Library of Congress

ISBN: 978-0-12-819491-1

For Information on all Elsevier publications
visit our website at https://www.elsevier.com/books-and-journals

Publisher: Matthew Deans
Acquisitions Editor: Edward Payne
Editorial Project Manager: Joshua Mearns
Production Project Manager: Sojan P. Pazhayattil
Cover Designer: Matthew Limbert

Contents

Preface *ix*

1. Hydrophilic and organophilic systems in nature **1**

1.1 Introduction 1
1.2 Prebiotic chemistry 2
1.2.1 Path from inorganic to organic matter is mediated through water 2
1.2.2 Conversion of inorganic materials to organic matter through a series of complex reactions 3
1.2.3 External sources of energy 4
1.3 Two-phase systems and amphiphilic molecules in living organisms 6
1.3.1 Cell membrane 6
1.3.2 Artificial photosynthesis 12
1.3.3 Exotic forms of matter 13
1.3.4 Nonaqueous environments for cell formation: azotosomes in cryogenic environments 13
1.4 Water and organic matter as interfaces in the formation of natural oil 16
1.5 Conclusions 17
References 18

2. Nature of the liquid–liquid interface **23**

2.1 Introduction 23
2.2 Electron and mass transfer in liquid–liquid interfaces 24
2.2.1 Mass transfer: diffusion occurring at an interface 24
2.2.2 Electron transfer: redox reaction at an interface 26
2.3 Interfacial tension at the boundary between two immiscible liquids 29
2.4 Supercritical fluids: fluids without surface tension and gas–liquid boundary 34
2.4.1 Supercritical carbon dioxide 34
2.4.2 Carbon dioxide as a solvent 34
2.5 Thickness of the liquid–liquid boundary 39
2.6 Conclusions 44
References 45
Further reading 48

3. Self-organization of nanomaterials at the interface **49**

3.1 Introduction 49
3.2 The formation of micelles 50
3.2.1 Catanionics: mixture of anionic and cationic surfactants 53
3.3 Organic—inorganic vesicles 54
3.4 Deoxyribonucleic acid—directed self-assembly 59
3.5 Interface energy and interfacial tension in thin films 60
3.6 Self-assembled thin films with plasmonic nanoparticles 67
3.7 Redox reactions from functionalized nanomaterials through liquid—liquid interfaces 72
3.8 Photonic crystals 75
3.9 Biomimetics, molecular recognition, and nanomachines at a liquid—liquid interface 76
3.10 Gyroid polymeric nanostructures from interfaces 78
3.11 Conclusion 79
References 80

4. Synthesis and generation of polymeric materials from interfaces **89**

4.1 Introduction 89
4.2 Plastic polymers synthesized through addition or condensation polymerizations 91
4.3 Oxidative polymerization of conductive polymers 96
4.3.1 Polyaniline: interfacial polymerization of aniline 96
4.4 Interfacial polymerization of pyrrole: formation of thin films and noble metal nanocomposites 100
4.5 Conductive polymer nanocomposites 101
4.5.1 Noble metal nanocomposites synthesized through reduction of metal ions by monomers 101
4.5.2 Carbon nanotubes and graphene nanocomposites 103
4.6 Applications and physical properties 104
4.6.1 Sensor applications 104
4.6.2 Single crystalline conductive polymers: the ordering of polyaniline chains 108
4.7 Electronic and magnetic properties 110
4.7.1 Transport mechanism: conductivity of polymers 110
4.7.2 Polyaniline nanoparticles with metallic electronic transport 111
4.7.3 Magnetic properties of conductive polymers 112
4.8 Other polymeric systems from interfaces 113
4.9 Conclusion 122
References 123

5. Light-emitting materials generated at the liquid–liquid interface 131

5.1 Introduction 131
5.2 Fundamentals of photoluminescence 134
5.2.1 Light emission 134
5.2.2 Intersystem crossing 136
5.2.3 Singlet oxygen generation 137
5.2.4 Fluorescence microscopy: applications at the oil–water interface 139
5.3 The superhydrophobic conductive polymers 141
5.3.1 Linear 3-perfluoroctyl polyaniline 143
5.3.2 Branched 4-perfluoroctyl aniline 146
5.3.3 Effect of substitution position on the physical properties. Cross-linked versus branched chains 151
5.4 Other photofunctional 2D materials 152
5.4.1 Metal–organic framework nanosheets through liquid–liquid or air–liquid interfaces 153
5.4.2 Covalent organic frameworks 154
5.5 Conclusions 155
References 156

6. Interfaces in industrial applications 161

6.1 Introduction 161
6.2 Energy applications: composite anode catalysts, self-assembled monolayers, and capacitive storage 161
6.3 Use of surfactants, polymers, nanoparticles, and supercritical fluids in oil and gas resources 166
6.3.1 Enhanced oil recovery with supercritical fluids 166
6.3.2 Role of asphaltenes in crude-oil emulsions 167
6.3.3 Cosolvents in lignin removal and cellulose treatment 171
6.4 Catalysts that operate in liquid–liquid interfaces and water-in-oil emulsions 172
6.5 Outlook and perspectives 182
References 183
Further reading 192

7. Outlook and perspectives 193

Abbreviations *195*
Index *197*

Preface

My interest in liquid–liquid interfaces, the reactions occurring at them and through them, as well as their physical properties and potential applications was initiated during my early days of doctoral studies at the National Center of Scientific Research "Demokritos." While I was working on the synthesis of polystyrene and polymethyl methacrylate composites with magnetic nanoparticles, I encountered the very interesting properties of conductive polymers and expanded my work on polyaniline-iron oxide nanocomposites. At this point, I became familiar with the work of Professor Richard Kaner on the interfacial polymerization of aniline for the synthesis of nanofibers with excellent size distribution homogenous shape, which results in their superior performance as sensors. This method was shown to be a versatile tool for the synthesis and design of nanocomposites with noble metal nanoparticles among other materials.

The book has six chapters and an extensive bibliography at the end of each chapter. Each of the six chapters is accompanied by an introduction and conclusion parts. Chapter 1, Hydrophilic and Organophilic Systems in Nature, initiates a discussion starting from the evolution of interfaces in natural environments. This discussion continues on the nature of the liquid–liquid boundary in Chapter 2, Nature of the Liquid–Liquid Interface, providing the physicochemical aspects (e.g., mass and electron transfer) behind this entity. This analysis on the nature of interfaces is necessary for understanding the processes that result in the selforganization of matter (which is the subject of Chapter 3, Self-organization of Nanomaterials at the Interface) and the synthesis of polymers and polymer thin films (Chapter 4: Synthesis and Generation of Polymeric Materials from Interfaces). Chapter 5, Light-Emitting Materials Generated at the Liquid–Liquid Interface, is a new chapter in this second edition and is dedicated on the fluorescent materials from interfaces. Finally Chapter 6, Interfaces in Industrial Applications, outlines the applications of interfaces and surfactants in catalysis, enhanced oil recovery with supercritical fluids, and other energy-related applications.

I would like to thank my publisher, Elsevier, Edward Payne and Joshua Mearns, for providing me with the opportunity to publish this updated second edition of the book. I hope that the direction of this work (i.e., fundamentals underlying the liquid–liquid interfacial systems and their applications) may be of interest in people working in industry, consultancies, as well as academia. I do not have "hands on" experience with all the techniques and methodologies that are outlined in this book so certain conclusions might be less than robust. Hence I would be very happy to receive comments and suggestions by readers.

Panagiotis Dallas
February 2020, Athens, Greece.

CHAPTER 1

Hydrophilic and organophilic systems in nature

1.1 Introduction

Water and carbon-based macromolecules are fundamental molecules and building blocks in living organisms and cells. The efficient self-assembly of the biomolecules through a series of delicate natural processes and their effective cooperation as a functional unit is a magnificent phenomenon that lies beyond our understanding and the repeated attempts of elucidation and reproduction in the laboratory. There are plenty of questions that need to be asked: how was this self-organization of materials initiated billions of years ago? Why are these specific molecules and atoms so abundant on Earth? Which physical processes enable the synthesis of complex molecules and what are the "rules" of self-assembly? We are all familiar with lipid membranes in cells and amphiphilic molecules, but can other nonaqueous systems and biomolecules in nonaqueous environments be models for self-organization of membranes?

Many of the constituents of these natural systems (e.g., small molecules, long-chain biomolecules, cells) are composed of asymmetric structures. However, mimicking these asymmetric formations in the laboratory has been difficult due to energy minimization effects. The first example of a particle self-assembling from smaller, individual, building blocks was shown by James W. McBain in 1913 [1,2]. This tendency for energy minimization leads to the preferential formation of lamellae or spherically shaped particles. These asymmetrical structures are very important and are maintained through complex systems and procedures.

Here an introduction to prebiotic chemistry is presented along with modern controversies on the nature and properties of nonaqueous environments that may be hosts for life forms. These have been speculations based on the abundance of nitrogen-based compounds on extraterrestrial planetary objects. Furthermore recent examples of light-harvesting functional nanomaterials that can influence and alter the ion flow through a lipid membrane are discussed in the context of newly developed active

Generation of Polymers and Nanomaterials at Liquid-Liquid Interfaces
DOI: https://doi.org/10.1016/B978-0-12-819491-1.00001-0

nanoparticles for biological applications. Besides the biological world there are many other natural systems of importance involving water—organic interfaces and binary systems. A discussion on the influence of water molecules on the formation of petroleum and natural gas resources during prolonged treatment under high pressure and high temperature in underground wells is outlined at the end of the chapter.

1.2 Prebiotic chemistry

1.2.1 Path from inorganic to organic matter is mediated through water

Our notion is that water is the fundamental molecule of life and that it is a necessary component for all known life forms and their sustainability [3]. Water molecules in solid, liquid, or gas forms are abundant on Earth. The total volume of water reaches the staggering level of 1.4 billion km^3, and the 97% of this amount is stored in oceans. Water was one of the primordial elements of Earth's environment and its atmosphere, along with carbon dioxide (CO_2), nitrogen, and hydrogen. There are questions regarding the source of water and a possible explanation is bombardment of our primordial and still-under-formation planet by extraterrestrial objects such as meteorites. This hypothesis is supported by various findings showing that some meteorites may contain $\leq$20% of water [4]. Also oxygen itself is one of the most abundant elements in the Earth's crust.

A comprehensive outline of the abundance of various elements in the earth's crust is shown in Fig. 1.1. Besides water living systems and life forms are based essentially on carbon (more specifically this carbon that is part of hydrocarbon molecules). These organic molecules are immiscible with water. Thus the functions of life forms are based on an interplay and cooperation between these systems. This cooperation involves an exchange of mass, electron charge, and energy between essentially two-phase systems. Since water is a highly polar solvent, this polarity enabled the spontaneous formation of "micromachines" such as membranes. The solvation of inorganic salts and minerals and their high heat capacity and heat of vaporization are crucial for stabilization of the temperature in the system. This stabilization and preservation of the chemical environment and metabolism in a closed living system is called "homeostasis."

The factors that were mentioned earlier have raised philosophical questions on why we need to continue seeking the existence of extraterrestrial life based on the presence of aqueous environments on other

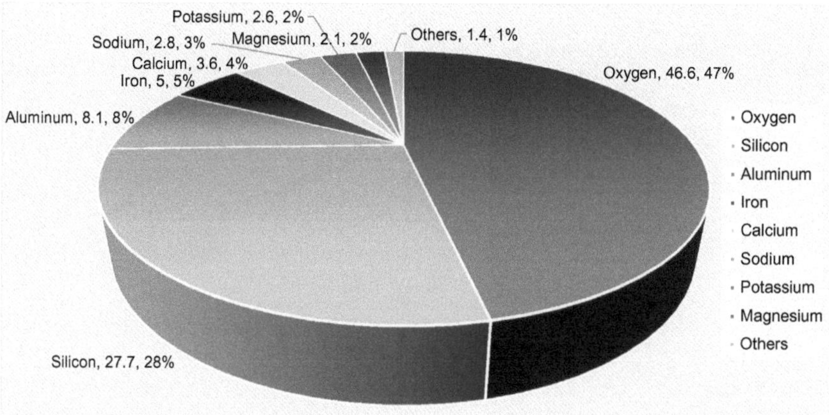

Figure 1.1 Percentage distribution of the most abundant elements in the Earth's crust. *J.E. Andrews, P. Brimblecombe, T.D. Jickells, P.S. Liss, An Introduction to Environmental Chemistry, Blackwell Science Ltd., Hoboken, NJ, 1996.*

planets. For example does the fact that life on Earth is based on water mean that water is the universal molecular of life? Can another type of molecule under different temperature, pressure, and chemical compositions have this role provided that "ideal" circumstances may be met? The main transitions in physical properties for a series of small and very common molecules that have been proposed to be candidates as alternative solvents compared with water (melting point, 0°C; boiling point, 100°C) are ammonia (−78°C, −33°C); methane (−182°C, −164°C); acetylene (−183°C, −89°C); and methanol (−94°C, 65°C). Finally a brief outline of extraterrestrial systems that are based on nitrogen-containing solvents is presented.

1.2.2 Conversion of inorganic materials to organic matter through a series of complex reactions

In the early environment of the Earth, the mixture of small molecules required heat, radiation, and energy to initiate the reactions between them that would lead to functional units and purpose-built organic materials. Surfactants are very important molecules in the formation of life through "bottom-up" transformation of small inorganic precursors due to their role in cellular assemblies. Fig. 1.2 shows a schematic representation of the speculative mechanism of cell formation. Surfactant assemblies and an interfacial environment can provide local reaction conditions different from those of a one-phase bulk system. They are expected to have

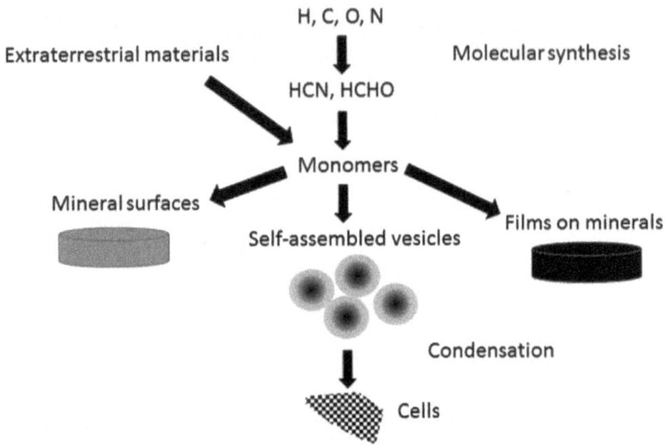

Figure 1.2 Scheme depicting the proposed synthesis of small molecules and biological biomolecules from inorganic matter and their subsequent self-organization into living cells.

different mechanical properties and may form helical, spherical, or elongated assemblies. A comprehensive review on these issues has been discussed thoroughly and presented by Peter Walde [5].

The energy needed for initiation of these physical, chemical, and biological phenomena was provided by the sun, the atmosphere, and the Earth. Solar radiation is dominated by visible light radiation, which is in the range 400–750 nm, with a maximum in the blue region (500 nm). Hence some of these solar photons are absorbed by plants, oceans, and the Earth, another part is reflected and, the final part is remitted as infrared electromagnetic radiation [6].

1.2.3 External sources of energy

The energy and heat needed for the initiation and propagation of chemical reactions in the primordial atmosphere were provided by atmospheric electric discharges, lightning, which can spontaneously produce intense heat, high-energy ultraviolet radiation, enormous current density, and an environment of reactive plasma. The primordial earth contained gaseous molecules such as methane, CO_2, water, ammonia, nitrogen, and hydrogen, and the electric discharge of the atmosphere acted as an intense source of ionizing radiation, light, and heat. These theories will remain speculation based on perhaps irrelevant data unless we can model, understand, and reproduce these processes in the laboratory, otherwise unambiguous historical evidence for their existence will be discovered.

Finally the Urey–Miller experiment is of fundamental importance for discussions on the generation of life. Urey and Miller carried out a classic experiment to demonstrate that the fundamental building blocks of living organisms, complex organic molecules, can be synthesized in the laboratory from smaller inorganic precursors in conditions that simulate the environment that dominated the Earth billions of years ago. They aimed to synthesize amino acids, which are building blocks of the proteins, the fundamental molecules for life. Their experiment was based on a reaction vessel that included the following molecular precursors: methane, ammonia, water, and hydrogen. Artificial lightning was generated from electrodes. Fig. 1.3 is a schematic representation of the experiment showing the yields of all the 20 basic amino acids that were obtained from the experiment [7], such as succinic acid (0.27%), formic acid (4%), glycine (2.1%), glycolic acid (1.9%), alanine (1.7%), sarcosine (0.25%), *N*-methyl alanine (0.07%), glutamic acid (0.051%), and lactic acid (1.6%). The initial experiment held great promise and could lead to uncovering the formation of life forms, but the limitation of this process became obvious immediately. Repeated experimental attempts revealed that no deoxyribonucleic acid (DNA) or ribonucleic acid (RNA) molecules could be obtained through the modified Urey–Miller experimental procedures tested, irrespective of the altered parameters. This methodology, despite its undoubtedly

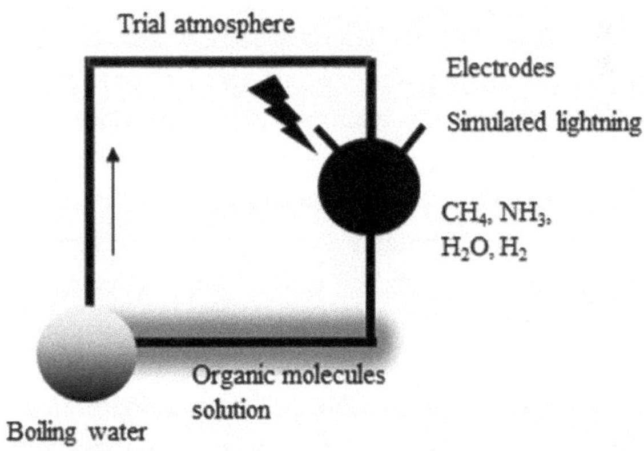

Figure 1.3 A schematic representation of the Urey–Miller experiment that resulted in the formation of amino acids in aqueous solution. In the text are presented the yields of some amino acids and small organic molecules that were obtained by the Urey–Miller experiment in 1952. Unfortunately the experiment was disappointing because molecules such as DNA or RNA were not observed. That is, self-organization toward functional matter is a more complicated process.

excellent insight and promising preliminary results, cannot be used to elucidate the formation, development, and reproduction of even the simplest organisms. Hence it is impractical for any theories, let alone experimental evidence, for cognitive human beings.

Undoubtedly cell formation requires the synthesis of very complex and "tailored" organic molecules and their delicate self-assembly in an aquatic environment with high salinity and ionic strength toward a functional "nanomachine." Furthermore there is strong evidence that the synthesis of amino acids in a primitive "molecular and ionic pool" took place through a mechanism known as the Strecker synthesis. It has been shown that Strecker synthesis type reactions did not take place during Urey–Miller experiments in their "synthetic pool." Small inorganic molecules such as HCN and HCO are formed from the dissociation of the precursors of nitrogen and water in arc plasma. Subsequently they react with NH_3 in aqueous solutions but this method does not provide stereochemical control over the carbon center.

1.3 Two-phase systems and amphiphilic molecules in living organisms

1.3.1 Cell membrane

The living cell as the fundamental "molecular machine" of the living organism will be discussed first. The cell has a water content of 70%. The outer cell membrane is the entity that separates it from the outside world and makes it the archetypical example of a two-phase system [7]. The biological cell membrane is a unique and functional nanostructure constructed through an ingenious self-assembly of phospholipids. The latter are densely packed, but can enable ion flow. Interestingly the mechanical properties of the membrane can be "tuned" and this elasticity is an important property that dictates cell function.

Thus depending on the phospholipid that constitutes the main building block of the membrane, it can undergo a sharp transition from a liquid-disordered state to solid-ordered phase in a characteristic transition temperature [8]. In addition the lipid bilayer assembly is a mobile entity and it can move across the membrane at $\approx 10^{-8}$ cm^2/s. In the presence of cholesterol this rate decreases because packing of phospholipids is induced. Various movements such as lateral, rotation, transbilayer are permitted.

The bilayer structure of the membrane was revealed first in 1925 through a series of experimental studies undertaken by Gorder and Grendel. After extraction of phospholipids from erythrocyte membranes, they added them to water and compressed. At this point they realized that the erythrocyte membranes now covered an area twice that of the cell membranes. The reason for this area coverage was due to the spontaneous arrangement of the phospholipids, with the polar group headed toward the aqueous phase and the organic hydrocarbon part headed toward air. When they were placed in water, the lipids disrupted the hydrogen bonding of the aqueous molecules (the so-called "hydrophobic effect").

In 1930 came the realization that biological membranes also contain proteins, thereby providing a more complicated image of the system. The cell membrane enables selective transport of ions through lipid arrays. This can be imagined as ion flow through flexible organic barriers. Control of ion channels through manipulation at the molecular level would be important to understand how neural networks work (and perhaps even be able to manipulate them). Many ideas and strategies have been proposed regarding manipulation of the membrane potential, including employing light- and stimuli-responsive functional nanomaterials [9,10].

As an example of materials that can control ion transport, Numata and coworkers used fullerenes, a class of electron–acceptor nanomaterials that were coupled with a ferrocene/porphyrin electron-donor moiety on the other side of a triad structure [11]. This resulted in an extremely large electric field of 10^6 V/cm, and a long-lived photoexcited, charge-separated state. Time-resolved fluorescence spectroscopy showed this lifetime to be 10 ns. Using this donor–acceptor system, they reported the photoinhibition of ion transport through a membrane and cell depolarization [11]. Control of ion transport and photoinduced depolarization was achieved through illumination of visible light at 400–450 nm. This spectral range was chosen because it lies at the optimum excitation wavelength of C_{60} fullerene. Analytic representation of the fullerene triad (with fullerene as the acceptor and ferrocene as the donor) and the process that is dictated by long-lived charge separation states is shown in Fig. 1.4.

Due to the amphiphilic nature of surfactants and their tendency to reduce the interfacial tension of liquid–liquid interfaces, various chemical reactions that may not take place under normal conditions can occur within a micelle or alternatively they can be accelerated to a large extent. This effect has been denoted as "micellar catalysis" [12–14]. Examples

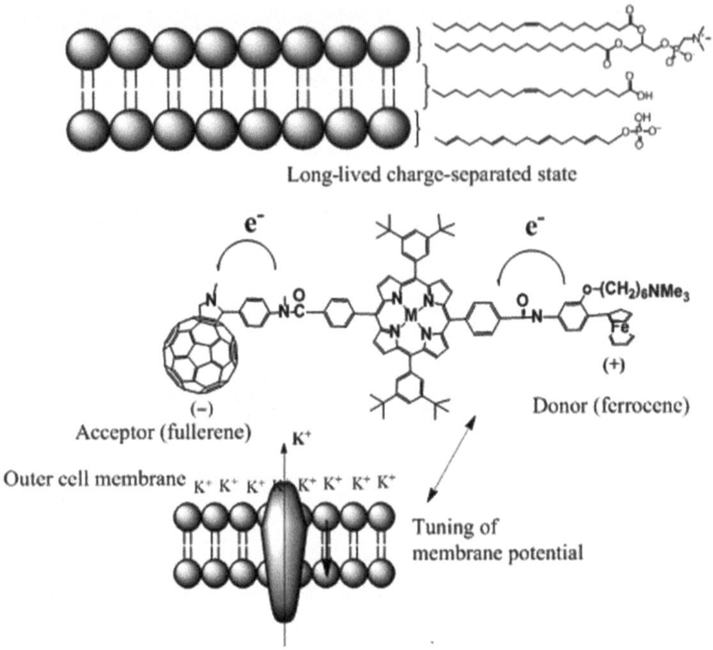

Figure 1.4 Schematic representation of the bilayer membrane and packing of phosphatidylcholine amphiphilic biomolecules. Harvesters of artificial light such as the complicated fullerene–macrocycle donor–acceptor triad generate high local electric fields and a long-lived photoexcited state. *Adapted from T. Numata, T. Murakami, F. Kawashima, N. Morone, J.E. Heuser, Y. Takano, et al., Utilization of photoinduced charge-separated state of donor–acceptor-linked molecules for regulation of cell membrane potential and ion transport, J. Am. Chem. Soc. 134 (2012) 6092.*

include organic reactions such as the Diels–Alder (DA) addition [15,16]. Also the molecular recognition between adenine and (thyminyloctyl) ammonium bromide can occur in a sodium dodecyl sulfate (SDS) micelle but cannot take place in the absence of the surfactant. An example of 1,3-dipolar cycloaddition mediated by micelles is depicted in Fig. 1.5. This reaction is accelerated by hexadecyltrimethylammonium bromide micelles, as demonstrated by Rispens and Engberts [18]. Other examples from the wealth of organic reactions are the Negishi cross-coupling reactions involving *E*- or *Z*-alkenyl substrates that take place in water with phase-transfer catalysis surfactants. Also a diester can be synthesized through coupling of a polyethylene glycol of molecular weight of 600 (α-tocopherol) and a naturally synthesized dicarboxylic acid (sebacic acid). In this example zinc with tetramethylethylenediamine, an electron-rich

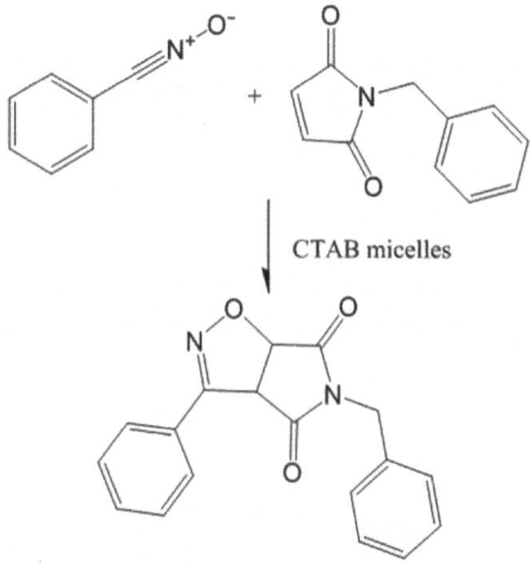

Figure 1.5 Example of an 1,3-dipolar cycloaddition reaction mediated by cetyltrimethylammonium bromide (CTAB) micelles. *For experimental details on the reaction, see A. Krasovskiy, C. Duplais, B.H. Lipshutz, Stereoselective negishi-like couplings between alkenyl and alkyl halides in water at room temperature, Org. Lett. 12 (2010) 4742 [17].*

palladium (II) catalyst, and the *E*- or *Z*-alkenyl halide and an alkyl halide partner at room temperature were used [19].

Another example of application of phase-transfer catalysis surfactants is the well-known Sonogashira coupling reaction that takes place between lipophilic alkynes and aryl bromides. This reaction leads to the efficient formation of products with avoidance of copper species under relatively mild conditions [20]. If the coupling reaction between phenyl acetylene and bromobenzene occurred in the presence of a phase-transfer catalyst surfactant, then a very good yield (83%) was achieved, whereas without addition of a surfactant a yield of only 34% was achieved. The authors completed the reaction among 20 combinations of terminal alkynes as well as electron-rich and electron-poor aryl bromides with excellent yields for the coupling reaction. Another example is the olefin metathesis reactions in aqueous environments [21]. Micellar catalysis in metathesis reactions has been provided by Lipshutz and coworkers [22,23].

1,3-Dipolar cycloaddition has common characteristics with DA addition reactions. A crucial difference is that, for 1,3-dipolar cycloadditions, the rate constants may decrease or increase depending on solvent polarity. However, in some cases, the rate constants are almost completely insensitive to the medium and are far less influenced by solvent polarity and the reaction environment compared with typical DA reactions and hardly exceed one order of magnitude over the full range of solvent polarities (i.e., from completely nonpolar *n*-hexane to very polar water molecules). Finally the progress of 1,3-dipolar cycloaddition and DA additions have been studied in polar media through a systematic study on the effects of micelles added to the system. The reaction rate is enhanced due to the higher concentration of reactants in the restricted micellar reaction volume serving as the reaction vessel. However, this effect is balanced by the micellar rate constant being at least one order of magnitude smaller than the rate constant in water. Finally the main factor determining these micellar rate constants was the effect of the local medium mimicked by a water/1-propanol mixture.

Based on the micellar contribution to base recognition, it has been proposed that the first cells on the Earth were lipid vesicles in which chemical reactions were facilitated and served in a confined, reactive chemical environment. The permeability of micelles (and consequently cell membranes) can also be modulated through temperature variations. Lower temperatures make the arrangement more solid-like and robust and do not let substances permeate, whereas higher temperatures decrease the order of arrangements and permeability increases. For an analytical presentation of catalytic reactions in micelles, see the review by La Sorella and coworkers (Fig. 1.6) [19].

The role of vesicles and micelles in biology has also been demonstrated through the interaction of proteins, which are polypeptide macromolecules with fluid lipid bilayers. This interaction leads to a change in the

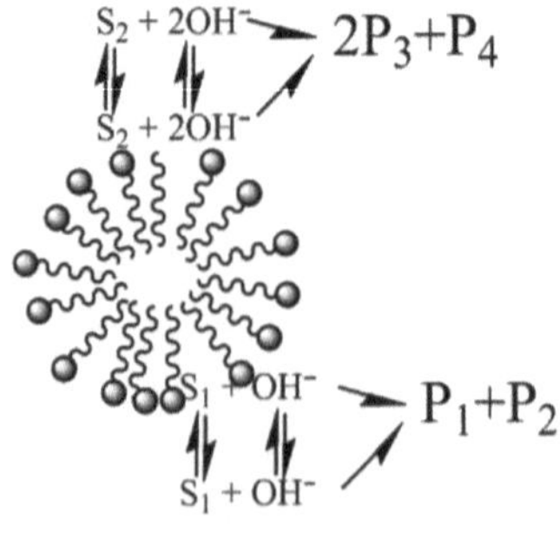

Figure 1.6 A micelle-driven reaction (S1: *p*-nitrophenylacetate; S2: fluorescein diacetate; P1: acetate, P2: *p*-nitrophenolate; P3: acetate; P4: fluorescein). *Adapted from T. Rispens, J.B.F.N. Engberts, Efficient catalysis of a Diels – Alder reaction by metallovesicles in aqueous solution, Org. Lett. 3 (2001) 941 and T. Namani, P. Walde, From decanoate micelles to decanoic acid/dodecylbenzenesulfonate vesicles, Langmuir 21 (2005) 6210 [24].*

conformation of the peptide and new functionalities. Furthermore laboratory experiments have shown that lipid bilayers can aid polymerization reactions and even synthesize RNA from polymerase enzymes [23]. Cationic or zwitterionic surfactants may be employed as vesicles [26–28]. Kuboi and coworkers studied the interactions between liposomes and denaturated proteins [28]. They considered the vesicle to be a different type of solvent that catalyzes the refolding of proteins. Fig. 1.7 is a schematic representation of these interactions. In a similar manner, Beck and coworkers prepared ordered hexagonal mesoporous materials using micellar systems and employing cationic surfactants with variable lengths of hydrocarbon chains [29]. The synthesized materials had large surface areas, similar to those of zeolites, and the role of surfactants in prebiotic environments and assembly of materials was indicated through these experiments.

Another effect of amphiphilic molecules and interfaces on the physicochemical processes of materials and molecules is an increase on the otherwise weak chemiluminescence from a Fenton-type reaction ($Co(II) + H_2O_2 + OH^-$) when the surfactant-like dodecyl benzene sulfonate (denoted DBS) intercalates layered double hydroxides (LDH) [30]. Fenton-type reactions are important in biology since the transition metal ions in the intracellular space lead to the formation of reactive free oxygen species such as superoxide, singlet oxygen, and hydroxyl radicals. This chemiluminescent enhancement of intercalated DBS was demonstrated

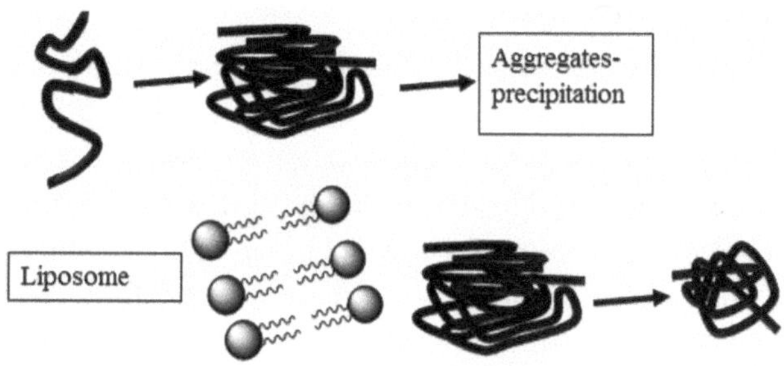

Figure 1.7 Folding of carbonic anhydrase enzyme is mediated and assisted by vesicles. After a 5 M solution of guanidinium hydrochloride is added to the enzyme solution, the enzyme transforms to its unfolded state. In the presence of a suspension containing vesicles, the refolding process goes through binding of the intermediate molten globular state to the outer surface of vesicles. *Adapted from R. Kuboi, M. Yoshimoto, P. Walde, P.L. Luisi, Refolding of carbonic anhydrase assisted by 1-palmitoyl-2-oleoyl-sn-glycero-3-phosphocholine liposomes. Biotechnol. Prog. 13 (1997) 828.*

through fluorescence spectroscopy, identification of radical scavengers, and electron microscopy. The DBS surfactant is relatively hydrophobic, however, it leads to the formation of the hydroxyl radical (•OH) through the reaction of cobalt cations with hydrogen peroxide. Subsequently •OH reacts within the LDH to form the sulfate radical $(\bullet SO_3^-)$. Recombination of $\bullet SO_3^-$ can lead to the formation of excited sulfur dioxide molecules SO_2^* with an emission maximum $(\lambda_{em}) = 430$ nm or, in other cases, of the excited singlet oxygen dimol species with $\lambda_{em} = 490-550$ nm. Electron spin resonance could also be used to detect •OH from intercalated LDH.

Regarding the time-resolved fluorescence measurements for estimating the excited states lifetime, a biexponential decay behavior has also been found earlier in individual aqueous micellar solutions of SDS, cetyltrimethylammonium bromide (CTAB), and Triton X-100. The decay timescale for the excited state is short (110–160 ps). Longer lifetimes have been reported for bulk short-chain alcohols whereas, in the case of cation–anion surfactant systems, both amphiphilic molecules are involved in construction of the interface [27]. Stokes shifts in catanionic vesicles have been found to be influenced by chain length, as demonstrated by studies on coumarin 153 emission dynamics [31]. For reactivity assisted by liposomes, see works on the polycondensation of amino acids and peptides within vesicles [32–35].

1.3.2 Artificial photosynthesis

"Natural" photosynthesis is the fundamental process used by plants to convert solar energy to chemical energy. This is achieved by utilizing electromagnetic radiation, water, and CO_2 through a series of very complicated processes. Enormous scientific activity has been dedicated for the development of new strategies inspired by natural plant photosynthesis in which compatible inorganic and biological entities can transform light, water, and CO_2 to higher-value chemicals. Examples of successful implementation of these strategies are formaldehyde, methane, or carboxylic acids. "Artificial" photosynthesis converts small molecules utilizing solar power. The energy demands of modern society are 20–40 terawatts (TW), and an ambitious target of obtaining 70 TW from solar energy by 2050 has been set [36d].

Recent works by Yang have demonstrated the importance of cells in the artificial photosynthesis. That is, the electrochemical reduction of CO_2 in an electrochemical cell. Fig. 1.8 shows a hybrid inorganic–biochemical synergetic

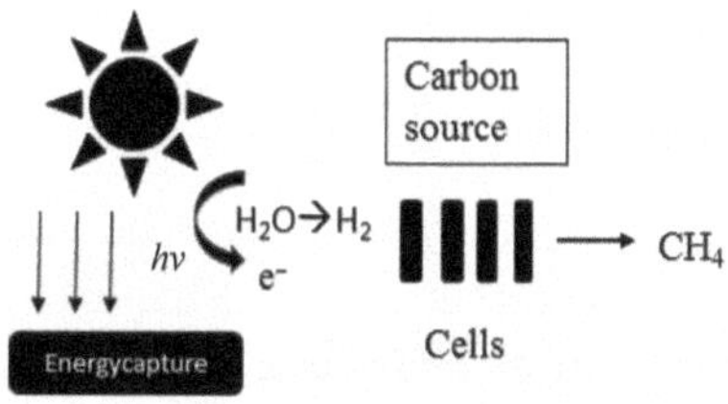

Figure 1.8 A hybrid approach for efficient photocatalysts for CO_2 reduction (schematic). *The diagram is based on the study by E.M. Nichols, J.J. Gallagher, C. Liu, Y. Su, J. Resasco, Y. Yu, et al., Hybrid bioinorganic approach to solar-to-chemical conversion, Proc. Natl. Acad. Sci. U S A 112 (2015) 11461; D. Kim, K.K. Sakimoto, D. Hong, P. Yang, Artificial photosynthesis for sustainable fuel and chemical production, Angew. Chem. Int. Ed. 54 (2015) 3259; C. Liu, N.P. Dasgupta, P. Yang, Semiconductor nanowires for artificial photosynthesis, Chem. Mater. 26 (2014) 415 [36].*

approach [36]. Examples of biological moieties are strains of *Ralstonia eutropha* for the aerobic production of isobutanol, isopropanol and 3-methyl-1-butanol, and cultures of *Methanosarcina barkeri*, which is an anaerobic archean [25].

1.3.3 Exotic forms of matter

Prebiotic environments were significantly harsher as well as more reactive and unpredictable compared with the present-day stable Earth and its atmosphere. The Urey–Miller experiment (arguably with some limited success) attempted to shed light on biomolecular formation (albeit achieving only the synthesis of small molecules) but other peculiar systems may be synthesized under extreme conditions. Electric discharges may generate a large amount of heat and plasma. Their application in specific systems may lead to the formation of exotic and uniquely symmetric materials.

Examples of such materials are carbon nanotubes and, most importantly, the fullerenes (the third allotrope of carbon after graphite and diamond). The high-vacuum arc discharge vaporization of graphite precursors leads to the synthesis of appreciable amounts of fullerenes, a family of highly symmetrical materials composed exclusively of sp^2 carbon atoms [37].

1.3.4 Nonaqueous environments for cell formation: azotosomes in cryogenic environments

The role of water as a fundamental molecule for life has been challenged and debated repeatedly. Despite intense efforts and huge astronomical projects, there is no proof for the existence of a living organism or a life outside the Earth. According to our perception, water is a requirement for the evolution of life because it comprises $\approx$70 wt.% of the mass of a cell. However, water in liquid form is rare in other planetary objects. Alternative solvation systems such as liquid methane, which is abundant, could be an alternative host environment and potential solvent medium, but cryogenic temperatures would be needed. In such conditions lipids and other hydrocarbon-based molecules would lose the mobility and flexibility they possess in the aquatic environments of the Earth.

Could self-organized vesicles, in a manner similar to liposomes, be formed and function in nonaquatic environments? In these environments, more extreme pressure and temperature conditions occur than those in present-day Earth.

To debate this concept further, Stevenson and coworkers demonstrated through theoretical studies that stable cryogenic membranes could be formed from molecules that have been observed in the atmosphere of Saturn's moon: Titan. The latter has an atmosphere composed of liquid methane close to its surface. According to the observation made by Cassini [38], in the upper atmosphere of Titan there are several nitrogen-containing small molecules in abundance. Stevenson and Clancy proposed that certain organic nitrogen compounds can function in liquid methane at very low temperatures [39]. Molecules resembling liposomes are now defined as "azotosomes" due to the presence of nitrogen (derived from the Greek word azoto). The authors employed molecular simulations to show that, in a cryogenic solvent, azotosomes may have elasticity equal to that of common lipid bilayers in water at room temperature.

It is the presence of liquid molecules in other planetary systems that sparked the interest in azotosomes. A prominent example will be the liquid methane solutions that can be an alternative model for a nonaqueous solvent. The azotosomes-based membranes have been predicted to have similar mechanical properties to those of conventional cell membranes. Fig. 1.9 shows a proposed aggregated assembly of azotosomes and a comparison between the interaction among the phospholipids and nitrogen-based surfactants in a liquid methane environment. The experiments with such systems require cryogenic temperatures. Hence till now studies can only be purely theoretical and are supported by extensive studies on molecular dynamics to demonstrate the crucial role of the nitrogen group. In recent work by Stevenson and coworkers, simulations showed that by comparing the azotosome-simulated membranes of hexane- and acetonitrile-based bilayers, in the absence of a nitrogen group, the layer appeared to be thick, brittle, and without flexibility, thereby demonstrating the importance of polar head groups. HCN could not self-assemble in an azotosome due to its very small size.

The abundance of various nitrogen-bearing compounds in the atmosphere of Titan are: HCN; 200 ppm, cyanoacetylene; 40 ppm, acrylonitrile; 10 ppm, cyanoallene; 4 ppm, acetonitrile; 3 ppm, 2,4-pentadiyne nitrile; 1 ppm, and propanenitrile; 0.5 ppm. These molecules and compounds may act as alternative solvents in extraterrestrial cryogenic conditions. The structure of Titan's atmosphere is shown in Fig. 1.10.

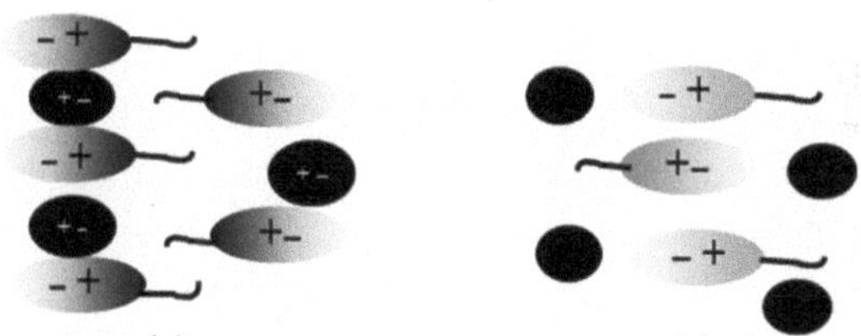

Figure 1.9 A vesicle is considered to have a diameter of 0.9 nm (90 Å). Representation of an assembly of liposomes in a polar solvent (A) and the azotosomes (B) in a nonpolar solvent. *A drawing of an azotosome vesicle can be seen in the work of J. Stevenson, J. Lunine, P. Clancy, Membrane alternatives in worlds without oxygen: creation of an azotosome, Sci. Adv. 1 (2015) e1400067.*

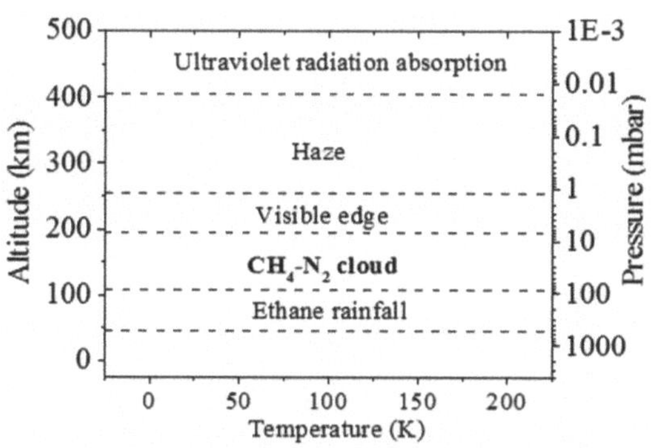

Figure 1.10 A diagrammatic representation of the atmosphere of Titan with respect to different pressures and temperatures. These observations are based on data from Voyager 1. *The diagram is based on the work by T. Owen (Nature 2005; 438: 756) and online sources.*

1.4 Water and organic matter as interfaces in the formation of natural oil

The role of two-phase systems in natural systems and environments is not restricted in the biological world: it is also valid in underground processes and the formation of natural resources. Examples include granite melts [41], coal formation [42,43], and clay mineral diagenesis [44]. Shimoyama and Johns provided further information on the catalytic cracking of fatty acids for hydrocarbon formation from petroleum [45]. Furthermore Whitney undertook hydrothermal tests on the reaction of smectite clays to illite transition. This transformation is a multistage process with various homogenous reactions. He used potassium-saturated montmorillonite at 250−400°C. These processes involve dissolution, solute mass transfer, and precipitation or crystallization. The latter is a complicated process that can include nucleation, crystal growth, Ostwald ripening, and crystal coalescence. The dissolution step is facilitated by cationic substitution within smectite layers. When the rock:water ratio was 20:1, the water within the lamellar space was the dominant medium of dissolution. According to previous works, the water molecules within the smectite layers are structurally different from bulk water, regardless of the overall amount of water in the system. The hydrocarbon small molecules and their amounts (micromoles) generated after heating behenic acid ($C_{21}H_{44}COOH$) at 100°C in the presence of 1 g of bentonite clay can be seen in detail in the work by Jurg and Eisma [45]. After heating for 89 and 720 h, respectively, the authors obtained the following yields for various hydrocarbons: propane (0.03 and 0.87 μmol); propene (0.15 and 0.25 μmol); *n*-butane (0.02 and 0.54 μmol); butanes (0.19 and 0.16 μmol); *n*-pentane (0.02 and 0.38 μmol); and *n*-hexane (0.02 and 0.29 μmol).

A water and hydrocarbon two-phase system has been proposed to mediate the formation of natural oil in underground environments. High pressures (≤3000 psi) and temperatures (> 100°C) have crucial roles in the formation of conventional and unconventional oil. Lewan showed that water is an important compound for CO_2 generation during pyrolysis under hydrous and anhydrous conditions. He demonstrated that significantly higher amounts of CO_2 are generated in the presence of water [46]. Lewan also demonstrated that water-saturated bitumen is responsible for oil formation, with different products being formed under hydrous or anhydrous conditions (Fig. 1.11) [48].

Oxygen originating from water is the source of the oxygen in the CO_2, but the remaining hydrogen can be used further for terminating

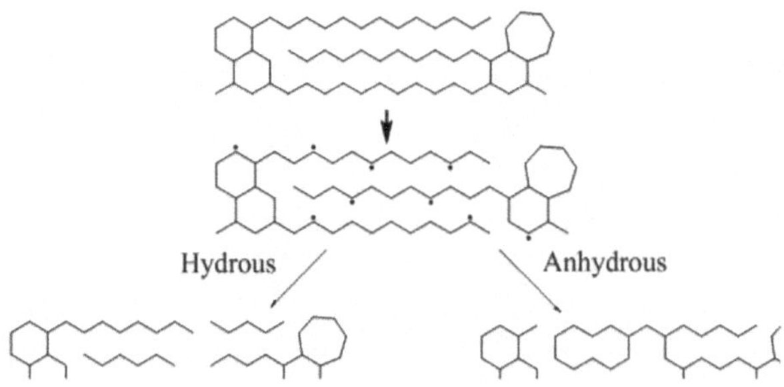

Figure 1.11 Two routes, the hydrous (320 g of water at 241 kPa of helium) and anhydrous, for decomposition through thermal maturation of bitumen (asphalt) after treatment of 72 h. The as-formed radicals are indicated with dots. *Adapted from M.D. Lewan, Abstr. Pap. Am. Chem. Soc. 204 (1992) 91.*

free-radical sites. Studies have shown catalytic conversion by clay minerals of organic molecules to petroleum hydrocarbons [46,48]. According to this hypothesis, water serves as the source of hydrogen for the formation of hydrocarbon molecules. For example on contact with water, a fatty, long-chain, organic acid, behenic acid generates smaller hydrocarbons after heating at 200°C in the presence of a clay mineral, bentonite. Lewan showed the small hydrocarbon molecules that are formed after heat treatment of the organic acid in the presence of bentonite clay [47]. The presence or absence of water influences the yields of the small hydrocarbons that are formed. As an example the ratio of isobutene or isopentane to normal butane or normal pentane is increased in that case.

During oil formation, normal alkanes with hydrocarbon chains of length 22–34 carbon atoms may also be formed. There is interplay between the amount of normal (*n*-) alkanes with an odd number of carbon atoms and those with an even number of carbon atoms in the range of 27–37 carbon atoms depending on the depth and timescale of burial because these parameters influence the pressure and temperature of the well. For example as the depth of the burial increases, the number of hydrocarbons with odd numbers of carbon atoms decreases and disappears gradually. Some theories suggest that these petroleum products have been synthesized from straight-chain fatty acids through decomposition of previous living matter.

1.5 Conclusions

Interfaces have important roles in biological, natural, and geological processes. Interfaces are fundamental for all organisms, cell membrane functionalities, and formation of underground oil in wells. There have been repeated efforts to replicate interfaces in man-made experiments. These processes required excessive heat, mass, and energy transfer whereas the self-organization of matter in naturally constructed systems is a delicate and high-precision procedure. Understanding, elucidating, and replicating the interfaces encountered in nature remain a challenge for physical sciences and materials engineering. Further advances in this field are expected to lead to more efficient energy-conversion devices, design of novel photocatalysts, and to inspiration in biomimetics.

Laboratory experiments have demonstrated the role of micellar assemblies and formations that can catalyze complex reactions through nanometer-scale entities that can be controlled simultaneously.

These are discussed in the last chapter of this book after giving a thorough introduction on the nature of the liquid—liquid interface, the rules of self-assembly, and the range of materials that have been synthesized through interfaces. Due to their fundamental roles in living organisms, the study of self-assembled systems has been discussed with philosophical questions regarding life initiation and life forms in alien environments with low temperatures, high pressures, and absence of water. These works have sparked interesting debates on various biogenesis theories, but we do not know how life originated. Even the simplest self-organization systems seem extremely difficult to visualize, let alone construct in the laboratory. Simultaneously they appear trivial and unimportant compared with the extremely complex and conscious human beings.

References

[1] J.W. McBain et al., Trans. Faraday Soc. 9 (2013), 93—107

[2] N.A. Kotov, Superstructures simplified, Nat. Nanotechnol. 11 (2016) 1002.

[3] J.E. Andrews, P. Brimblecombe, T.D. Jickells, P.S. Liss, An Introduction to Environmental Chemistry, Blackwell Science Ltd., Hoboken, NJ, 1996.

[4] (a) M.A. Sephton, Organic compounds in carbonaceous meteorites, Nat. Products Rep. 19 (2002) 292.
(b) A.R. Sarafian, S.G. Nielsen, H.R. Marschall, F.M. McCubbin, B.D. Monteleone, Early accretion of water in the inner solar system from a carbonaceous chondrite—like source, Science 346 (2014) 623.

[5] P. Walde, Orig. Life Evol. Biosph. 36 (2006) 109.

[6] G.E. Thomas, Cambridge Atmospheric and Space Sciences Series Radiative Transfer in Atmosphere, Cambridge University Press, Cambridge, 2008.

[7] S.R. Goodman, Medical Cell Biology, second ed., Lippincott-Raven, Philadelphia, PA, 1998.

[8] B.M. Stott, M.P. Vu, C.O. McLemore, M.S. Lund, E. Gibbons, T.J. Brueseke, et al., Use of fluorescence to determine the effects of cholesterol on lipid behavior in sphingomyelin liposomes and erythrocyte membranes, J. Lipid Res. 49 (2008) 1202.
[9] (a) S. Szobota, E.Y. Isacoff, Optical control of neuronal activity, Annu. Rev. Biophys. 39 (2010) 329.
(b) M.D. Molin, Q. Verolet, A. Colom, R. Letrun, E. Derivery, M. Gonzalez-Gaitan, et al., Fluorescent flippers for mechanosensitive membrane probes, J. Am. Chem. Soc. 137 (2015) 568.
[10] H. Imahori, Y. Sekiguchi, Y. Kashiwagi, T. Sato, Y. Araki, O. Ito, et al., Long-lived charge-separated state generated in a ferrocene–meso, meso-linked porphyrin trimer–fullerene pentad with a high quantum yield, Chem. A Eur. J. 10 (2004) 3184.
[11] T. Numata, T. Murakami, F. Kawashima, N. Morone, J.E. Heuser, Y. Takano, et al., Utilization of photoinduced charge-separated state of donor–acceptor-linked molecules for regulation of cell membrane potential and ion transport, J. Am. Chem. Soc. 134 (2012) 6092.
[12] J.F. Rathman, Micellar catalysis, Curr. Opin. Colloid Interf. Sci. 1 (1996) 514.
[13] G. Li, Y. Liu, W. Xu, A. Song, J. Hao, Transition of phase structures in mixtures of lysine and fatty acids, J. Phys. Chem. B 118 (2014) 14843.
[14] (a) A. Roy, M. Maiti, S. Roy, Spontaneous formation of vesicles by sodium 2-dodecylnicotinate in water, Langmuir 28 (2012) 12696.
(b) J. Čejková, F. Štěpánek, M.M. Hanczyc, Evaporation-induced pattern formation of decanol droplets, Langmuir 32 (2016) 4800.
(c) F. Pratama, G.J. Blanchard, A comparison of energy flow in micelle and vesicle structures, J. Phys. Chem. B 119 (2015) 3025.
[15] D.C. Rideout, R. Breslow, Hydrophobic acceleration of Diels-Alder reactions, J. Am. Chem. Soc. 102 (1980) 7816.
[16] (a) M.B. Gawande, V.D.B. Bonifácio, R. Luque, P.S. Branco, R.S. Varma, Benign by design: catalyst-free in-water, on-water green chemical methodologies in organic synthesis, Chem. Soc. Rev. 42 (2013) 5522.
(b) M.-O. Simon, C.-J. Li, Green chemistry oriented organic synthesis in water, Chem. Soc. Rev. 41 (2012) 1415.
[17] A. Krasovskiy, C. Duplais, B.H. Lipshutz, Stereoselective negishi-like couplings between alkenyl and alkyl halides in water at room temperature, Org. Lett. 12 (2010) 4742.
[18] T. Rispens, J.B.F.N. Engberts, Efficient catalysis of a Diels–Alder reaction by metallo-vesicles in aqueous solution, Org. Lett. 3 (2001) 941.
[19] G. La Sorella, G. Strukul, A. Scarco, Recent advances in catalysis in micellar media, Green Chem. 17 (2015) 644.
[20] B.H. Lipshutz, D.W. Chung, B. Rich, Sonogashira couplings of aryl bromides: room temperature, water only, no copper, Org. Lett. 10 (2008) 3793.
[21] (a) J. Tomasek, J. Schatz, Olefin metathesis in aqueous media, Green Chem. 15 (2013) 2317.
(b) Skowerski, G. Szczepaniak, C. Wierzbicka, è. Gułajski, M. Bieniek, K. Grela, Highly active catalysts for olefins metathesis in water, Catal. Sci. Technol. 2 (2012) 2424.
[22] B.H. Lipshutz, G.T. Aguinaldo, S. Ghorai, K. Voigtritter, Olefin cross-metathesis reactions at room temperature using the nonionic amphiphile "PTS": just add water, Org. Lett. 10 (2008) 1325.
[23] B.H. Lipshutz, S. Ghorai, G.T. Aguinaldo, Ring-closing metathesis at room temperature within nanometer micelles using water as the only solvent, Adv. Synth. Catal. 350 (2008) 953.

[24] T. Namani, P. Walde, From decanoate micelles to decanoic acid/dodecylbenzenesulfonate vesicles, Langmuir 21 (2005) 6210.

[25] (a) D. Deamer, A giant step towards artificial life, Trends Biotechnol. 23 (2005) 336.
(b) J. Peretó, P. López-García, D. Moreira, Ancestral lipid biosynthesis and early membrane evolution, Trends Biochem. Sci. 29 (2004) 469.

[26] K.J. Hallock, D.K. Lee, J. Omnaas, H.L. Mosberg, A. Ramamoorthy, Membrane composition determines Pardaxin's mechanism of lipid bilayer disruption, Biophys. J. 83 (2002) 1004.

[27] T.J. McIntosh, The 2004 biophysical society-avanti award in lipids address: roles of bilayer structure and elastic properties in peptide localization in membranes, Chem. Phys. Lipids 130 (2004) 83.

[28] R. Kuboi, M. Yoshimoto, P. Walde, P.L. Luisi, Refolding of carbonic anhydrase assisted by 1-palmitoyl-2-oleoyl-sn-glycero-3-phosphocholine liposomes, Biotechnol. Prog. 13 (1997) 828.

[29] (a) J.S. Beck, J.C. Vartuli, W.J. Roth, M.E. Leonowicz, C.T. Kresge, K.D. Schmitt, et al., A new family of mesoporous molecular sieves prepared with liquid crystal templates, J. Am. Chem. Soc. 114 (1992) 10834.
(b) E. D'Aguanno, E. Altamura, F. Mavelli, A. Fahr, P. Stano, P.L. Luisi, Physical routes to primitive cells: an experimental model based on the spontaneous entrapment of enzymes inside micrometer-sized liposomes, Life (Basel) 5 (2015) 969.
(c) T. Pereira de Souza, F. Steiniger, P. Stano, A. Fahl, P.L. Luisi, Spontaneous crowding of ribosomes and proteins inside vesicles: a possible mechanism for the origin of cell metabolism, ChemBioChem 12 (2011) 2325.
(d) P.A. Monnard, D.W. Deamer, Membrane self-assembly processes: steps toward the first cellular life, Anat. Rec. 268 (2002) 196.
(e) D.W. Deamer, The first living systems: a bioenergetic perspective, Microbiol. Mol. Biol. Rev. 61 (1997) 239.

[30] L. Zhang, Z. Zhang, C. Lu, J.M. Lin, Improved chemiluminescence in Fenton-like reaction via dodecylbenzene-sulfonate-intercalated layered double hydroxides, J. Phys. Chem. C. 116 (2012) 14711.

[31] N. Sarma, J.M. Borah, S. Mahiuddin, H. Al Rasid Gazi, B. Guchhait, R. Biswas, Influence of chain length of alcohols on stokes' shift dynamics in catanionic vesicles, J. Phys. Chem. B 115 (2011) 9040.

[32] A. Grochmal, L. Prout, R. Makin-Taylor, R. Prohens, S. Tomas, Sequence-specific, nanomolar peptide binding via cucurbit[8]uril-induced folding and inclusion of neighboring side chains, J. Am. Chem. Soc. 137 (2015) 12269.

[33] M. Blocher, D. Liu, P. Walde, P.L. Luisi, Liposome-assisted selective polycondensation of α-amino acids and peptides, Macromolecules 32 (1999) 7332.

[34] M. Blocher, D. Liu, P.L. Luisi, Liposome-assisted selective polycondensation of α-amino acids and peptides: the case of charged liposomes, Macromolecules 33 (2000) 5787.

[35] T. Hitz, M. Blocher, P. Walde, P.L. Luisi, Stereoselectivity aspects in the condensation of racemic NCA − amino acids in the presence and absence of liposomes, Macromolecules 34 (2001) 2443.

[36] (a) E.M. Nichols, J.J. Gallagher, C. Liu, Y. Su, J. Resasco, Y. Yu, et al., Hybrid bioinorganic approach to solar-to-chemical conversion, Proc. Natl. Acad. Sci. U S A 112 (2015) 11461.
(b) D. Kim, K.K. Sakimoto, D. Hong, P. Yang, Artificial photosynthesis for sustainable fuel and chemical production, Angew. Chem. Int. Ed. 54 (2015) 3259.
(c) C. Liu, N.P. Dasgupta, P. Yang, Semiconductor nanowires for artificial photosynthesis, Chem. Mater. 26 (2014) 415.

(d) N.M. Haegel, et al., Terawatt-scale photovoltaics: Transform global energy, Science 364 (2019) 836.
[37] W. Kratschmer, L.D. Lamb, K. Fostiropoulos, D.R. Huffman, Solid C60: a new form of carbon, Nature 347 (1990) 354.
[38] C. McKay, H. Smith, Possibilities for methanogenic life in liquid methane on the surface of Titan, Icarus 178 (2005) 274.
[39] J. Stevenson, J. Lunine, P. Clancy, Membrane alternatives in worlds without oxygen: Creation of an azotosome, Sci. Adv. 1 (2015) e1400067.
[40] N.L. Bowen, O.F. Tuttle, Am. Mineralogist 37 (1952) 286.
[41] J.P. Schuhmacher, F.J. Huntjens, D.W. van Krevelen, Fuel 39 (1960) 223.
[42] E. Berl, A. Schmidt, Ann. Chem. 493 (1932) 97.
[43] G. Whitney, Clays Clay Miner. 38 (1990) 343.
[44] J.W. Jurg, E. Eisma, Petroleum hydrocarbons: generation from fatty acid, Science 144 (1964) 1451.
[45] A. Shimoyama, W.D. Johns, Catalytic conversion of fatty acids to petroleum-like paraffins and their maturation, Nat. Phys. Sci. 232 (1971) 140.
[46] M.D. Lewan, Evaluation of petroleum generation by hydrous pyrolysis experimentation, Philos. Trans. R. Soc. Lond. A 315 (1985) 123.
[47] M.D. Lewan, Abstr. Pap. Am. Chem. Soc. 204 (1992) 91.
[48] M.D. Lewan, J.C. Winters, J.H. McDonald, Generation of oil-like pyrolyzates from organic-rich shales, Science 203 (1979) 897.

CHAPTER 2

Nature of the liquid–liquid interface

2.1 Introduction

This chapter discusses the nature, thickness, polarity, and shape of a liquid–liquid interface. This can be seen as a first step before further elaborating in subsequent chapters on its applications on the self-organization of nanomaterials, synthesis of polymers, use of supercritical fluids for mixing with oil or water, and materials for energy storage catalysis and biomimetics. This is of particular importance because ions, electrons and, in general, mass transfer can take place between two separate phases with a rate and selectivity different from their one phase, liquid, counterparts. This process may be "tuned" selectively with regard to the nature of the organic medium (i.e., its polarity, interfacial tension with water, and density) a series of novel architectures may be generated. Hence two different solvents that comprise the liquid–liquid interface will dictate the structure and properties of the materials that are synthesized.

The study of these phenomena is very important with respect to the synthesis of new materials and phase transfer processes through interfaces and because many organic compounds are disposed of in the aqueous environment and form separate entities with water. The transfer of electrons and mass between the two immiscible liquids are discussed in this chapter. Also the early experiments with biological systems are presented along with the Markus theory of electron transfer and Fick's model for mass transfer.

In this chapter, it is important to provide an introduction on the properties and characteristics of supercritical fluids and their interfaces with aqueous solutions. Thus the properties of supercritical fluids, an important and unique class of fluids without surface tension and with many industrial and fundamental research applications, are outlined at the end of the chapter. A rigorous analysis of their applications, especially with respect to their microemulsions with perfluorinated or hydrocarbon surfactants and aqueous systems, is presented in Chapter 4, Synthesis and Generation of Polymeric Materials From Interfaces, and Chapter 6, Interfaces in Industrial Applications.

Generation of Polymers and Nanomaterials at Liquid-Liquid Interfaces
DOI: https://doi.org/10.1016/B978-0-12-819491-1.00002-2

2.2 Electron and mass transfer in liquid–liquid interfaces

2.2.1 Mass transfer: diffusion occurring at an interface

Mass transfer through a liquid–liquid interface is a typical process in biological systems and polymerization experiments conducted in laboratories. These are systems in which reagents exchange between two phases including intramolecular charge transfer [1–9]. Polymerization reactions using mass diffusion are analyzed in Chapter 4, Synthesis and Generation of Polymeric Materials From Interfaces, in great depth because a wealth of important polymers can be synthesized. In this discussion, the interface is treated as a membrane that permits the transfer of both ions and molecules. It is well known that if a membrane separates two solutions with different concentrations, denoted as C_1 and C_2 (e.g., we will take into account that $C_1 > C_2$), flux takes place from the solution with a high concentration (C_1) to the solution with a low concentration (C_2) by following Fick's first law of diffusion (Fig. 2.1). The latter states that the net rate of diffusion "J" obeys the following law:

$$J = -D \times A \times \frac{\mathrm{d}c}{\mathrm{d}x} \tag{2.1}$$

where A is the surface area of the membrane and D is the diffusion coefficient. The diffusion coefficient represents the rate at which the molecule can pass through the membrane and is described by the Einstein equation (Eq. 2.2):

$$D = K \times \frac{T}{6} \times \pi \times r \times \eta \tag{2.2}$$

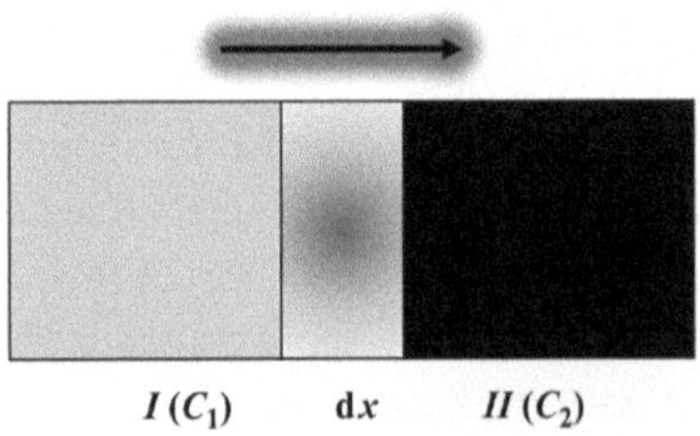

Figure 2.1 Example of diffusion through a membrane that permeates solute ions or molecules but not solvent molecules to pass toward one of the two separated phases. In this specific example the concentration gradient is $C_1 > C_2$.

where K is the Boltzmann constant ($1.38064852 \times 10^{-23}$ m^2 kg/(s^2 K)), T is temperature, r is the molecular radius, and η is the viscosity of the medium.

Fick's simplified equation is applicable to ideal systems with small molecules that bear no charge since charged ions are causing an electric field that influences the diffusion [10a]. The following representation will be presented to clarify the two Adolf Fick's laws that describe the diffusion flux and predict how the diffusion causes the concentration to change with time:

$$J = -D\left(\frac{\partial \Phi}{\partial \chi}\right)$$

which is the first Fick's law.

$$\left(\frac{\partial \Phi}{\partial t}\right) = D\left(\frac{\partial^2 U}{\partial x^2}\right)$$

which is the second Fick's law.

In more complicated systems, diffusion is dependent on the concentration gradient and electrochemical gradient (i.e., potential) on the interface. Fick's first law has been used to study the growth of inorganic nanoparticles (NPs) in solution whereby the diffusion of monomers to the interface dictates their growth and their size and shape. In this case Fick's first law must be rewritten to take into account the distance between the NP to the bulk concentration of ions in solution using Eq. (2.3):

$$J = \left[\frac{4\pi Dr(r+\delta)}{\delta}\right](C_b - C_i) \tag{2.3}$$

where δ is the distance, C_b and C_i are the concentration of monomers in solution and the solid–liquid interface, respectively, and C_r is the solubility of the particle [10b]. Fig. 2.2 shows the diffusion layer structure of the nanocrystal and the monomer concentration as a function of the distance (with the shaded area denoting the diffusion layer).

A liquid–liquid interface enables the molecular recognition between various molecules to take place in biological systems. These types of interactions might take place in supercritical fluids as well [11]. To model molecular recognition in biological media, the hydrogen bonding interactions in an interface have received extensive attention. The formation of riboflavin complexes with diamino-*s*-triazine in an aqueous–oil interface has been shown to be mediated through the formation of triple hydrogen

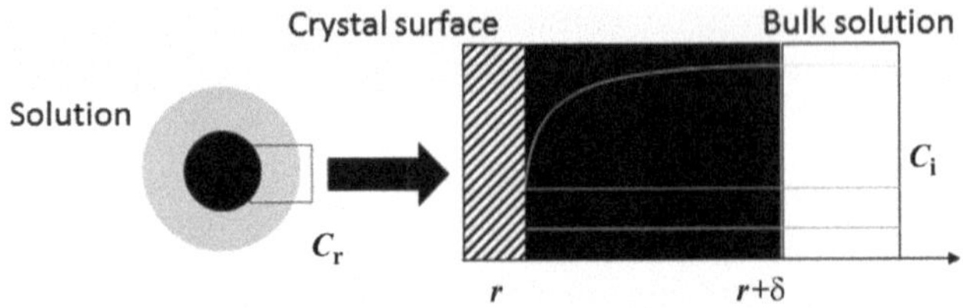

Figure 2.2 The diffusion layer near the surface of a nanoparticle [10].

bonding [12]. Also Nochi and coworkers showed that alkali metal cations such us Li^+ and K^+ form complexes with crown ethers at a water–heptane interfacial system [12,13a]. Sutjiadi-Sia and coworkers reported mass transfer phenomena in various solvent systems, including water, ethanol, corn germ oil, and carbon dioxide (CO_2) [13d]. They observed unique drop-related phenomena such as "drop kicking," upward motion of the continuous phase, and rising bubbles related to CO_2. They also reported the contact angle of ethanol and water at pressures up to 27 MPa in CO_2 on substrates such as Teflon, stainless steel, and glass.

2.2.2 Electron transfer: redox reaction at an interface

Besides the mass transfer of reagents, it is feasible to achieve electron transfer that results in current flow through a liquid–liquid interface. This current flow is dependent on the potential difference between the two solutions. In addition this current is (as expected) proportional to the number of the charged electrons that are transferred across the interfacial membrane. With respect to these current gradients, energy conversion reactions can occur at biological membranes. In the following paragraph, the basic principles behind the Marcus theory of electron transfer, a theory that has been applied closely to biological systems are outlined briefly.

In optical spectroscopy, the Franck–Condon principle is used widely to explain maxima in absorbance and fluorescence spectra based on the coupling between ground and excited electronic and vibrational states. This principle was developed and applied in the 1950s as a potential source for the faster reaction rates observed between large ions such as $Fe(CN)_6^{3-}$, $Fe(CN)_6^{4-}$, MnO_4^-, and MnO_4^{2-}. These reaction rates are faster than those involving smaller ions. The mechanism of this electron transfer was elucidated by Rudolph Marcus. He was awarded the Nobel Prize for Chemistry in 1992 for pioneering studies of electron transfer reactions. A brief schematic representation of the mechanism in biological systems is shown in Fig. 2.3. Markus formulated the mechanism of the electron transfer reaction by expressing the rate constant for the reaction in conjunction with the Gibb's free energy of activation using the "re-organization energy" (λ). The latter is the energy related to molecular rearrangement. In particular his theory has been applauded and extensively utilized by biochemists. Detailed studies on photosynthetic system II (photosystem II) demonstrated electron transfer from chlorophyll to a "special pair" ($BChl_2$) and subsequently to a pheophytin. These processes are ultrafast (2 ps) electron transfers.

Various typical electron transfer reactions through interfaces were studied by Samec and coworkers [6]. They demonstrated an electron transfer

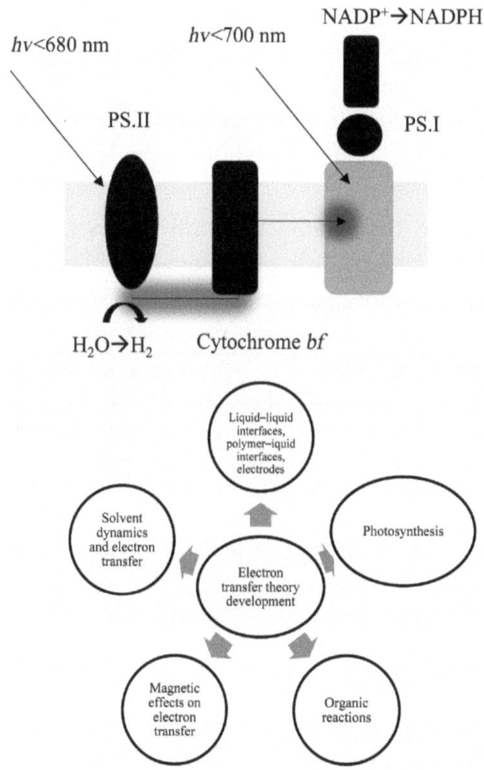

Figure 2.3 The study of electron transfer in chemical reactions was pioneered by Rudolph Marcus. The diagram is a simplified version of the one shown in his Nobel Prize lecture. It is a schematic representation of the electron transfer reactions taking place in photosystem II (NADP stands for nicotinamide adenine dinucleotide phosphate). This figure is not complete and a full picture can be found in the relevant literature.

between ferrocene molecules that are dissolved in nitrobenzene (nb) and an aqueous solution of $Fe(CN)_6^{3-}$ using the following mechanism:

$$Fc(nb) + Fe(CN)_6^{3-}(water) \leftrightarrow Fc^{+}(nb) + Fe(CN)_6^{4-}(water)$$

The redox reactions mediated through immiscible liquids can be classified into two distinct groups depending on the presence or absence of an external electromagnetic field. The first group consists of the spontaneous chemical reactions that occur in the absence of an external electromagnetic field. This type of reactions can be seen in processes such as bioenergetics, model membrane systems, and nanocrystalline thin films (which are grown and subsequently stabilized at oil—water interfaces) [7]. The redox reactions that belong to the second group [8] occur at the interface due to the presence of external electrical fields (e.g., those generated during a cyclic voltammetry experiment). The second type of interfacial electron transfers is, in most cases, a multistage process. The steps that are usually encountered involve diffusion of reactants to the interface, where they are absorbed; then the products formed move away or toward one of the two phases of a film.

An example from the biological world comes from the chlorophyll, which is a broad term for a class of important, photosensitive, molecules. The chlorophylls play a fundamental role in photosynthesis. They catalyze electron transfer reactions at oil—water interfaces and on bilayer lipid membranes [9]. Rabinowitch and Weiss observed a redox reaction of chlorophyll (Chl) with ferric chloride, for chlorophyll *a* and *b* [8]:

$$Chl + Fe^{3+} \rightarrow oChl^{+} + Fe^{2+}$$

where oChl is oxychlorophyll (a yellow oxidized form).

The redox reaction involves the addition of iron cations, whereas the chlorophyll molecule is dissolved in methanol solution. After the redox reaction between iron and chlorophyll, there is a visible color change from green to yellow. In general the high catalytic activity of *chlorophyll α molecules* in a two-phase system is due to a "hydrophilic head" stemming from the four pyrrole rings that constitute the conjugated part with carbonyls around a magnesium ion and a hydrophobic aliphatic chain that decorates it. Another example of phase transfer is the controlled electrodeposition of nanoparticles at the surface of an emulsion droplet [10a]. Gold nanostructures could be deposited through electrochemical methodologies that involve the reduction of aqueous phase Au(III). This is achieved through heterogeneous electron transfer from decamethylferrocene present in an emulsion droplet and the nanoparticles growth is taking place at the interface. It was found that the deposition of gold resulted in an increased current response for the single electron transfer of decamethylferrocene and cryo-TEM provided the imaging of the gold nanostructures. The reduction of gold leads to the build-up of negative ions within the aqueous phase and positive ions within the organic phase and a neutral charge of the interface is maintained during the through transfer of ClO_4^- ions between the two phases.

Cyclic voltammetry measurements took place after an emulsion was formed.

It is well demonstrated that the high surface tension between two immiscible liquids leads to an energetic benefit for the assembly of nanostructures. This will reduce the area in contact and the electrochemical control over the deposition of metallic structures at a liquid–liquid interface lead to fabrication of a four-electrode system that enables a potential directly to the interface. These studies may ultimately lead to a bottom up construction of artificial biological and/or nonbiological counterparts through a careful combination. This will produce microscale assemblies with biomimetic properties and behavior [10b,c].

Examples in this aspect are the liposome stabilized Pickering emulsions of phase-separated droplets that can be dextran-rich phase within polyethyleneglycol-rich phase. These phases are stabilized by liposomes or a liposome assembly around coacervate droplets comprised spermine [10d].

2.3 Interfacial tension at the boundary between two immiscible liquids

Due to the nonfavorable coordination of atoms on the surface of a material, to a large extent, the surface forces are those who dictate the properties of a material. The surface atoms or molecules are surrounded by molecules and atoms having different characteristics and behavior compared with those of interior molecules. Different interactions, including forces between particles, and between the particles and the medium, may take place in a colloidal system [11]. We can expand the concept of the air–solid interface to the variety of possible liquid–liquid interfaces. In that case the collective interaction of forces would give rise to the interfacial tension. The mean field theory created by Fisk and Widom explained the behavior and features of liquid–liquid and liquid–vapor interfaces, and they predicted that the surface tension would depend on the density profile of the two liquids among other parameters [14]. Immiscible high energy liquid–liquid interfaces may be classified in the order of increasing interfacial tension: water–oil < water–silicone < water–fluorocarbon. The interfacial tension, "γ," by definition is the partial derivative of the internal energy "U" with respect to the area "A" and is dependent on the physical conditions and the chemical composition that can be expressed from the following Eqs. (2.4) and (2.5):

$$\gamma = \left(\frac{\partial U}{\partial A}\right) S, V, n \tag{2.4}$$

$$S \times dT - V \times dP + A \times d\gamma + \Sigma n_i \times d\mu_i = 0 \tag{2.5}$$

where S is the entropy, T the temperature, V the volume, p is the pressure, n the mole amount of component, and μ the chemical potential.

In the case that one immiscible liquid is dispersed within the other then the system will form an emulsion. The dispersed phase is in the form of globules that are dispersed inside the continuous phase. Depending on the nature of globules, we will encounter the following cases: water-in-oil, oil-in-water, and multiple emulsions (see Fig. 2.4 for the first two cases). The latter case is a complex system in which water–oil and oil–water interfaces are dispersed within another immiscible liquid phase. The types mentioned are encountered routinely in polymer and materials chemistry, pharmaceutical, cosmetics, and petroleum sectors.

Fowkes demonstrated that the surface tension should be considered to be a sum of independent terms with each representing a particular intermolecular force [15] or, in different terms, as the excess of energy due to molecular or atomic configurations at an interface. Water forms strong

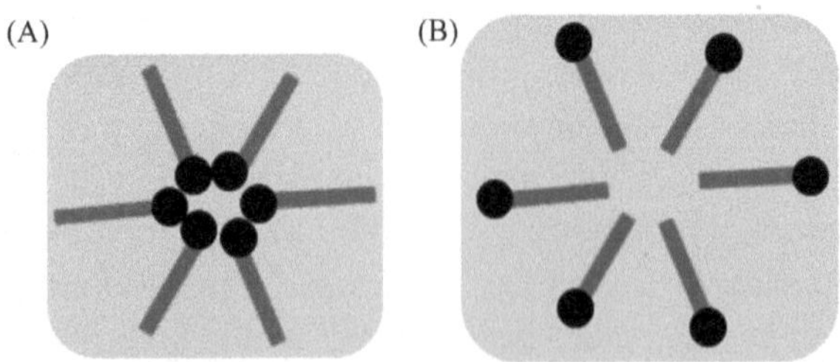

Figure 2.4 The arrangement of a surfactant is water-in-oil (A) and oil-in-water (B) types of emulsions. The hydrophilic part is colored black and the hydrophobic part is light gray.

intermolecular bonds, such as hydrogen bonding "*h*," and dispersion forces "*d*," that can be described by Eq. (2.6):

$$\gamma_w = \gamma_{wh} + \gamma_{wd} \quad (2.6)$$

Such strong intermolecular bonds are in contrast with organic molecules, particularly hydrocarbons "γ_c" that have a surface tension which is dependent exclusively on dispersion forces. Consequently for a description of the forces at a water–hydrocarbon interface, the Good–Girifalco equation, Eq. (2.7) may be applicable [16]:

$$\gamma_{cw} = \gamma_c + \gamma_w - 2\Phi(\gamma_c\gamma_{wd})^{1/2}$$

Φ is the interfacial interaction parameter.

With respect to the previous discussion, the interfacial organization of matter may lead to the construction of arrays and particles with anisotropic shapes. An interesting example of particle transformation through interfaces was published by Park and Furst, and a schematic representation of their work is shown in Fig. 2.5 [16]. As an aqueous phase, they used a 2 wt.% gellan solution, which initially is dissolved at 90°C and then becomes a viscous gel if cooled to less than 30°C–35°C. For the organic phase they used decane and sulfonated polystyrene microparticles were diffused through an isopropanol solution at the interface. The surface charge of the sulfonated polystyrene particles was measured to be

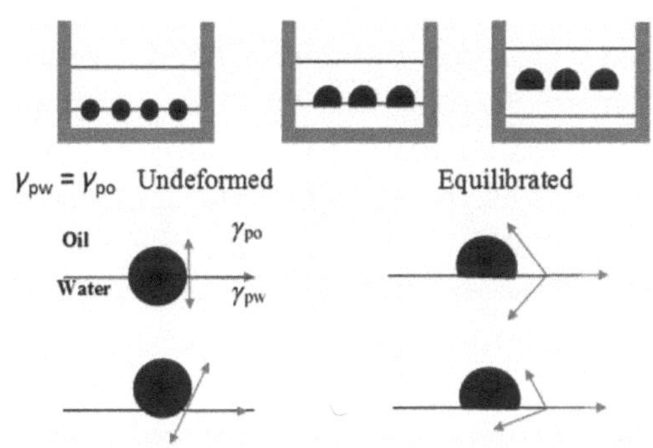

Figure 2.5 An oil–water interface was utilized by Park and Fust for the synthesis of anisotropically shaped polystyrene nanoparticles. The difference in interfacial tension between the water (aqueous solution with 2 wt.% gellan)–decane interface, along with further effects from PS, led to shape transformation and unusual disc-like microparticles. *Adapted from B.J. Park, E.M. Furst, Fabrication of unusual asymmetric colloids at an oil – water interface, Langmuir 26 (2010) 10406. Copyright: American Chemical Society.*

7.4 μC/cm^2 and no surfactants were added to the system. After an incubation period they removed the organic solvent (decane) and observed an unusual disc-shaped formation of particles using scanning electron microscopy. The particles appeared to be flat on the side exposed to the aqueous phase and had a curvature on the side of the organic phase. They calculated that the mass of each particle is conserved during the entire interfacial diffusion process. Hence it is a purely transformative process and not an exchange of PS macromolecules among different particles. They also concluded that a higher initial dissolution temperature for the gellan–water system accelerates particle transformation.

Various methods have been developed in the last decades for measurement of the interfacial tension among various liquid–liquid interfaces. The most widely employed methods have been the following: ring method (the du Noüy ring method is a technique for measuring the surface tension of a liquid); plate; rod; drop volume; spinning drop; pendant drop. In the ring method, a wettable ring is used for the measurement of the interface force as it moves from one phase to the other. During the measurements taking place through the plate method, a plate is immersed vertically in the lower phase whereas, in the rod method, a cylindrical rod is used. For the drop method the volume of a drop at a vertical capillary in the second liquid is measured. Finally the pendant drop method is based on the shape of a drop on a needle in a bulk liquid phase. For the solid–liquid interface the contact angle is used as the optimum method for measurement of the surface tension.

Other methods used for the measurement of interfacial tension include use of "micro channels" and the "electrowetting" methodology [17a,b]. Lee and coworkers developed an alternative method based on ionic liquids, with 1-alkyl-3-methylimidazolium methyl sulfate and 1-alkyl-3-methylimidazolium ethylsulfate that were used as test samples [17c]. The ionic liquid bearing the larger, ethylsulfate, chain demonstrated a lower interfacial tension compared with the ionic liquid with a small methyl sulfate chain. The net charge on the droplet did not change the shape of the droplet. Consequently deformation of the ionic liquid droplet took place under an applied external electric field of 3 kV/cm. The deformation D is the difference between the length of the two axes, the major a, and the minor axis b of the droplet, and this is divided by the sum of the two (Eq. 2.8). The authors believe that their newly developed methodology would be particularly useful for highly viscous fluids.

$$D = \frac{(a-b)}{(a+b)} = \left(\frac{9}{16}\right)\left[R_0 \times \varepsilon_{\text{out}} \times \frac{E^2}{\gamma_{\text{int}}}\right] \tag{2.8}$$

where R_0 is the radius of a nondeformed droplet, ε_{out} is the permittivity of the dielectric liquid, and γ_{int} is the interfacial tension between the two liquids.

The interfacial tension may be controlled along with the shape and properties of the interface upon surfactant addition. These surfactants would tend to arrange in a way that minimizes the interfacial area per unit volume, thereby following Eq. (2.9):

$$Av = d\varphi - 1 \tag{2.9}$$

where φ is the volume fraction of the hydrophobic chains, and d is the dimensionality, where $d = 1$ for planar interfaces, $d = 2$ for cylindrical, and $d = 3$ for spherical domains.

An extensive analysis of the behavior of surfactants will be presented in Chapter 3, Self-Organization of Nanomaterials at the Interface. In general their packing and arrangement within the solvent depend on their shape with a dimensionless parameter, the packing parameter "g" (defined by the ratio of the volume "V," the mean curvature "H," the Gaussian curvature "K" and the effective head group area "a_c") and the critical length "L_c" according to Eqs. (2.10)–(2.12). Fig. 2.6 is a schematic representation of these parameters and for details see the work by Antonietti and Forster [18]. As a general rule aggregation of amphiphilic molecules is entropically unfavorable, so the molecules need enthalpic stabilization. Detailed studies of interfaces, amphiphiles, and forces between liquid boundaries have been done over many years by the research team led by Jacob Israelachvili.

$$g = 1 + H \times L_c + K \times \frac{L^2}{3} \tag{2.10}$$

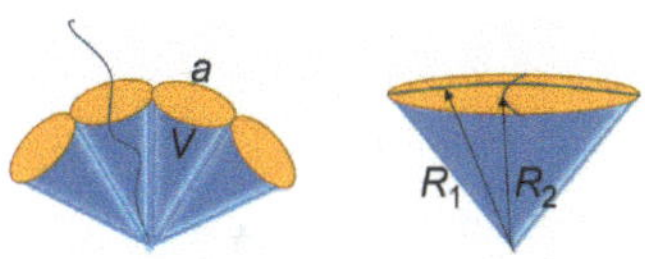

Figure 2.6 The different shapes that may be adhered to by the surfactants and the parameters that determine it. *Adapted from M. Antonietti, S. Forster, Vesicles and liposomes: a self-assembly principle beyond lipids, Adv. Mater. 15 (2003) 1323.*

$$H = \frac{1}{2}\left(\frac{1}{R_1} + \frac{1}{R_2}\right) \tag{2.11}$$

$$K = \frac{1}{R_1 \times R_2} \tag{2.12}$$

2.4 Supercritical fluids: fluids without surface tension and gas–liquid boundary

2.4.1 Supercritical carbon dioxide

While the oil–water system is the most commonly studied, there are other interfaces that must be considered besides the classical water–hydrocarbon interfaces. Examples include the carbon dioxide–water and fluorocarbon–water interfaces, both of which present many opportunities for energy-related and superhydrophobic coating applications. Carbon dioxide is a versatile, nontoxic, nonflammable, and low-cost green solvent that offers an environmentally friendly solution compared with the toxic and flammable organic solvents. Hence initially we must pay special attention to the nature of carbon dioxide as a quadrupole molecule that can act as a solvent as also on the behavior of the perfluorinated carbons (a class of materials with unique solubility behavior). CO_2 is a nonpolar solvent with low dielectric constant and zero dipole moment. It has found applications in refining, extraction, creation of polymer foams, supercritical drying, and cleaning of materials with a low dielectric constant [16]. It possesses a quadrupole moment because it is a charge-separated molecule, and so can act as a Lewis acid and Lewis base [17]. In general chemists consider it to be a similar solvent to dioxane. Water–carbon dioxide microemulsion systems are important in biology [18,19], organic synthesis [20], nanoscience and chemistry [13], and dry cleaning [12].

2.4.2 Carbon dioxide as a solvent

CO_2 has a low critical temperature (31.1°C) and pressure (1072 psi). These parameters make the carbon dioxide in its supercritical state particularly attractive for a series of applications such as the flooding of underground oil wells; supercritical drying and supercritical gels that lead to porous materials compared with other supercritical fluids. In general exactly at a critical point the compressibility of the fluid reaches infinity, thereby it is obeying the following Eq. (2.13):

$$\left(\frac{\partial P}{\partial V}\right)T = 0;\ \left(\frac{\partial^2 P}{\partial^2 V}\right)T = 0 \tag{2.13}$$

Supercritical fluids are a unique category of fluids and solvents due to the absence of surface tension (there is no liquid–gas boundary) and a "tunable" density and viscosity depending on the temperature and pressure. A comprehensive analysis of the application of supercritical CO_2 in harsh environments is presented in Chapter 6, Interfaces in Industrial Applications. These applications are mostly focusing on oil recovery with the use of carbon dioxide in its supercritical phase.

Examples of the phase transitions of CO_2 and noble gases in liquid or solid state indicates that different bonds will be formed under suitable circumstances. For example, xenon can have a phase transition to solid and has an enthalpy of melting of 14.9 kJ/mol. These weak short-range bonds arising from an uneven local distribution of electro density are termed "van der Waals interactions or bonds." Dipoles are formed instantly due to differences in the electron density in a molecule. Infrared spectroscopy has demonstrated that CO_2 can form dimers or even trimers [21]. Fig. 2.7 shows the dipolar structures of water and CO_2 and formation of oligomers between monomers of the latter.

Carbon dioxide has been investigated for use as a green, nonflammable, and abundant low-cost solvent. Its miscibility with water and oil at various temperatures and pressures has been the subject of intense investigation with respect to the use of supercritical CO_2 as a flood fluid in oil-recovery applications. To that end several research efforts have been devoted in developing surfactants to stabilize water–CO_2 emulsions. Carbon dioxide seems to have a preference toward fluorocarbons, whereas hydrocarbons are not soluble in liquid carbon dioxide or supercritical carbon dioxide [22]. This peculiar preference for interaction between carbon dioxide and the fluorine atom is based on theoretical studies and density-dependent ^{1}H- and ^{19}F-nuclear magnetic resonance spectroscopy. The explanations of the nature of the interactions that lead ultimately to the very high solubility of fluorocarbons in supercritical carbon dioxide are varying [23,24]. To address this question, Dalvi and Rossky undertook extensive atomistic molecular dynamics simulations. They started their simulations from *n*-octane and, upon stepwise addition of fluorine atoms

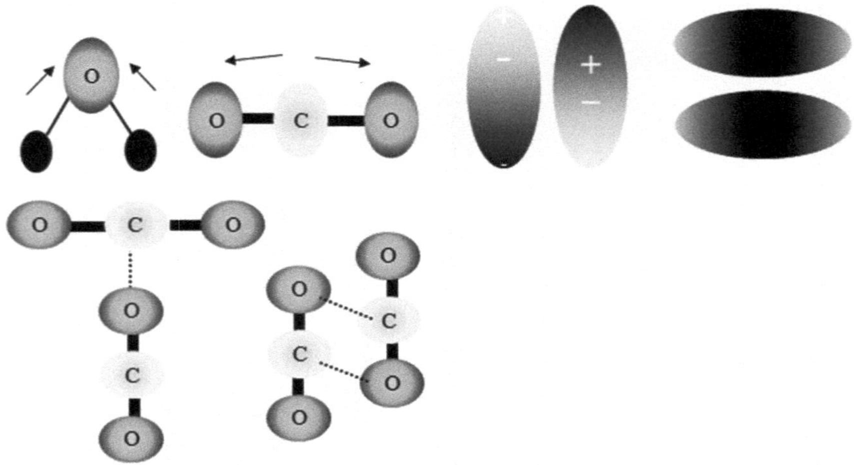

Figure 2.7 Various interactions among water and CO_2 molecules and optimized configuration of CO_2 dimers. Different charge distribution among a dipole and a quadrupole molecule is also shown.

they concluded their studies with *n*-perfluorooctane. A schematic representation of the thermodynamic cycles involving the transitions between gas and soluble states in water was shown by Dalvi and Rossky [25]. They calculated the difference in the hydration free energies of perfluorooctane and octane 2.54 kcal/mol, and assigned the greater hydrophobicity of fluorocarbons to size effects.

Saraji and coworkers revealed the physical properties and interfacial tension for carbon dioxide–water systems (Fig. 2.8 and Table 2.1) on a quartz surface in various conditions in subcritical and supercritical regions [26a]. This strategy is very important for carbon dioxide sequestration because it involves mixtures of supercritical carbon dioxide with rock formations, brine solutions, and crude oil. In their work, "wettability" was defined through measurements of receding and advancing contact angles. When the carbon dioxide bubble is growing, thereby having an increased contact with quartz, it leads to water recession from the quartz surface. This is called the "CO_2-advancing angle" or "water-receding angle" depending on from which phase it is measured. The values for the density of water and carbon dioxide, interfacial tension, and its respective

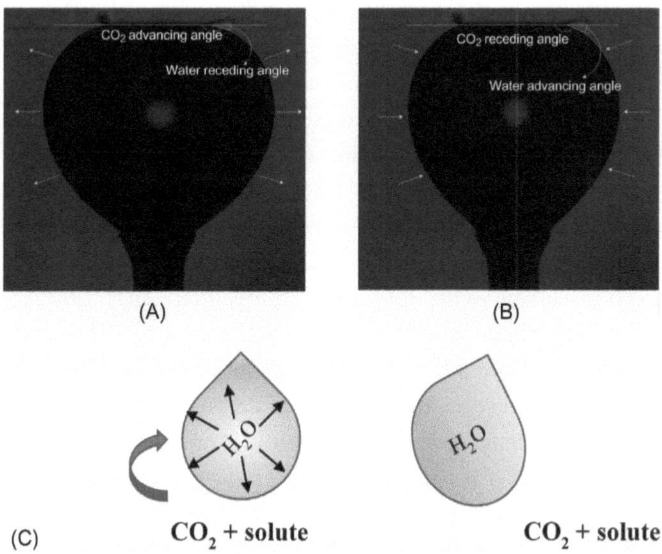

Figure 2.8 (A and B) Bubbles of supercritical CO_2 in water at 1700 psi and 45°C and (C) the behavior of water–CO_2–solute systems. *(A and B) Reproduced with permission from S. Saraji, L. Goual, M. Piri, H. Plancher, Langmuir 29 (2013) 6856; © American Chemical Society. (C) Based on the work from Y. Sutjiadi-Sia, P. Jaeger, R. Eggers, J. Supercrit. Fluids 46 (2008) 272. Copyright: Elsevier.*

Table 2.1 The density and interfacial tension in a carbon dioxide--water–quartz mixture at different pressures and temperatures.

Pressure (psi)	Temperature (°C)	ρ_{CO2} (kg/m³)	ρ_{H2O} (kg/m³)	$\Delta\rho$	Interfacial tension
1700	60	412.9	997.2	584.3	31.16
1300	60	233.4	995.4	762	31.98
900	60	130.1	992.3	862.2	42.31
500	60	62.5	986.6	926.1	50.78
1700	35	762.2	1011.5	249.3	32.29
1300	35	661.3	1010.4	349.2	32.84
900	35	170.1	1007.9	837.8	38.26

Source: Reproduced from S. Saraji, L. Goual, M. Piri, H. Plancher. Langmuir 29 (2013), 6856. © American Chemical Society.

temperature and pressure are shown in Table 2.1. They demonstrated that quartz may become less wetted by water as the pressure and temperature increase which, in a field test, may result in lower capillary pressures and carbon dioxide removal. Also Bikkina detailed contact-angle measurements of a CO_2–water–quartz system at 200–3000 psi and 25°C–50°C [26b]. Recently Chen and coworkers undertook molecular dynamics simulations to investigate the dependence of the pressure and temperature of the contact angle in silica–carbon dioxide–water systems [26c]. Their work demonstrated that the dependence of pressure and temperature is controlled by the surface groups of silica.

Fluorocarbons are considered to be the most efficient carbon dioxide thickeners. However, recent studies have revealed an anomalous solubility of acetate-based molecules in liquid carbon dioxide and in supercritical carbon dioxide, albeit with lower solubility than that observed with fluorocarbons. Hydrogen bonding between the hydroxyl groups and the carbon dioxide has also been indicated by shifts in Fourier transform infrared bands. Other molecules that can act as carbon dioxide thickeners are the sugar acetates. This is a class of compounds highly soluble in supercritical carbon dioxide. For sugar acetates the typical cloud points in supercritical carbon dioxide are at 120 MPa and 298K for a concentration of 5 wt.% [27]. Besides carbon dioxide other molecules possess a quadrupole moment, for example, chlorine, carbon monoxide, nitrogen, and boron fluoride. Values of the quadrupole moment can be estimated through application of an electric field gradient in a low-density gas to induce birefringence. This parameter was calculated by Graham and coworkers, who were the first to report a quadrupole moment for boron trifluoride [28a]. A good source of information on the polar behavior of small molecules is the work by Bundgen and coworkers [28b]. They undertook studies ab initio of the dipole moment and quadrupole moment (values in $\times 10^{-40}$ C m^2) of: O_2, NO, and N_2 (−4.67); CO (−9.47); HF, HCI, N_2O, and CO_2 (−14.27); OCS, CS_2, NH_3, ClH_2, O_3, SO_2, CH_2O, C_2H_4, and Cl_2 (10); BF_3 (−9.47) [28b]. Values were calculated by Graham and coworkers from birefringence measurements induced in a gas by an applied electric field [28a].

Various methods have been employed for determination of the polarity of the interface. With regard to spectroscopic methods, time-resolved total internal reflection fluorescence has revealed information on the environments around 8-anilino-1-naphthalenesulfonate at an interface between water and heptane. Data obtained by this method revealed that the interfacial polarity ($P_{A/B}$) was between that of heptane and water [2]. Other studies undertaken by Ishizaka and Kitamura [3] and Wang and coworkers [4] demonstrated that the $P_{A/B}$ had a value that was the average polarity of each phase (P_A and P_B). These conclusions were found to be valid for interfacial systems such as water–*n*-heptane, water–*n*-decane, and water–cyclohexane.

Selection of the organic medium will dictate the growth and formation of polymers or the arrangement of NP at an interface that can follow different pathways depending on the nature of the interface. In many cases, for example, polypyrrole, the product of the reaction is a complete polymer or composite insoluble in any solvent and exhibits extensive crosslinking. Thus as the film grows, it stays at the interface and can readily be isolated from there. In several other cases such as functionalized NP or polyaniline (PANI), the products move to the organic or aqueous phase, with charged species moving in the aqueous phase and neutral species in the organic phase. This phenomenon has been observed repeatedly for polyelectrolyte-like polymers such as PANI. Their oxidation states, the presence of surfactants acting as counterbalancing anions, and protonation dictate the outcome of the process.

2.5 Thickness of the liquid–liquid boundary

In general the interface is considered to be an inhomogeneous entity. Mathematical consideration of the interface according to Gibb's model of capillary forces depicts the interface as a two-dimensional (2D) entity.

With regard to the shape of the liquid–liquid boundary, one must take into account the thermal- and concentration-induced fluctuations that lead to a nonhomogeneous entity. Hansen [29] considered the interface as a three-dimensional (3D) region because, despite being very thin, it can be regarded as a non-2D entity. In Hansen's approach, an alternative convention was proposed in which two Lagrange multipliers were introduced to have values whereby the coefficients of the two chemical potential terms were eliminated. The derivative representing the pressure coefficient of the interfacial tension $(\partial\gamma/\partial P)_T$ was assigned to be the thickness of an interface, the Hansen thickness "τ." However, the derivative that leads to τ is not the physical thickness of the interface but is only the distance between two zero mass planes [29].

In the classical theory of capillary forces, the interface is a 2D mathematical surface, an inhomogeneous region with a thickness of a 1–2 nm (a negligible value compared with the bulk material) [30a]. The force on the interface, the interfacial tension, is caused by excess energy, entropy, and inhomogeneous mass distribution [30b]. However, there are many difficulties in theoretical studies and measurement methods so, in effect, the interfacial composition and thickness cannot be measured directly. Sometimes it must be treated as a uniform, 3D phase in the framework of macroscopic thermodynamics (Fig. 2.9). Most studies have dealt with an approximate estimation of the order of the boundary thickness derived from molecular theories. A numerical analysis on the withdrawal transition between two immiscible liquids that have similar viscosities was carried out by Berkenbusch and coworkers. They demonstrated that withdrawal causes deformation to the interface, from several shapes to a single shape formed at complete removal [30c]. Many other works have employed the pressure dependence of the interfacial tension to measure the interfacial thickness, following methods developed by the "Brussels school" [30d].

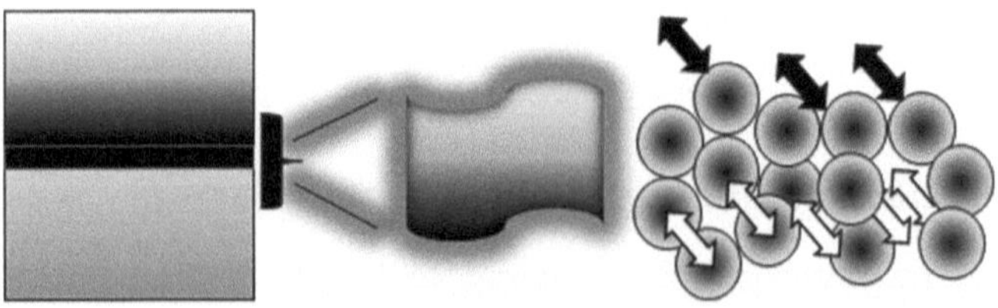

Figure 2.9 Competition between intramolecular (*white arrows*) and surface tension (*black arrows*) forces in a liquid–liquid interface.

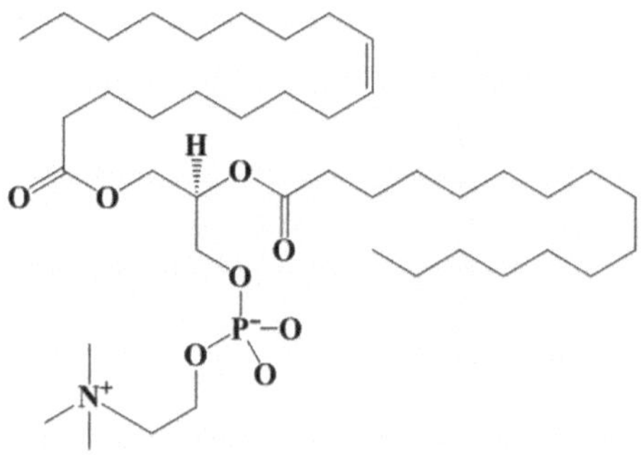

Figure 2.10 Molecular structure of a typical lecithin molecule bearing nonpolar aliphatic chains and a polar P—O group.

Formation of droplets that are not distributed uniformly in the interface and which migrate to both phases has been observed under light microscopy. Hence aggregation of water molecules takes place at the organic phase, and aggregation of hydrophobic molecules takes place in the aqueous phase.

Repulsion may take place among ions or molecules, and lecithin—water interactions have been used as models for two-phase systems (Fig. 2.9). Lecithins are a broad class of well-known amphiphilic, fatty macromolecules that may be found in animal and plant tissues and have been extensively used as emulsifiers in food chemistry. Regular sources of lecithin are meat, canola, cottonseed, soybean, and eggs. A typical structure of a lecithin molecule is shown in Fig. 2.10. The repulsion between lecithin bilayers has been shown to be particularly strong [30e]. In lecithin—water systems it appears that ≤ 21 water molecules per lecithin molecule are oriented on it.

Despite the fundamental importance of the CO_2—water surfactant for industrial- and energy-related applications, molecular understanding of the behavior of surfactants at the CO_2—water interface is insufficient [31]. Stone and coworkers undertook extensive computer simulations comparing two surfactants: one with a double-tail phosphate surfactant with a perfluorinated carbon tail, $Na^+\ {}^-O_2P(OCH_2CH_2(CF_2)_5CF_3)_2$, and a second one with similar structure but without fluorine, just a typical hydrocarbon chain, that is, $Na^+\ {}^-O_2P(O(CH_2)_7CH_3)_2$. They demonstrated that (1) CO_2 is a good solvent for the hydrophobic tails of both surfactants and (2) water molecules penetrate hydrocarbon surfactants to a greater extent than fluorocarbon surfactants. They concluded that the worse performance of hydrocarbon surfactants compared with perfluorinated surfactants was they could not separate the water and CO_2 phases from each other effectively. The structure of the interface is more complicated than it appears due to statistical fluctuations: thermal and density [32]. Even in the absence of external stimuli and motion, the interface between two fluids appears to be smooth, but thermal motions lead to a rough structure [33]. Thermal fluctuation in interfaces was predicted in 1908 by von Smoluchowski [34]. In subsequent works at the beginning of the twentieth century, other researchers provided a quantitative description of interface roughness in terms of thermally excited capillary waves [35] as revealed by light scattering [36] and X-ray scattering [37] experiments. Displacement of the interface can be described quantitatively and qualitatively through Eq. (2.14) with regard to the equipartition theorem:

$$\langle |h_k|^2 \rangle = \left(\frac{k_B T}{\gamma L^2}\right)\left(\frac{1}{k^2 + \xi^{-2}}\right) \tag{2.14}$$

where h_k defines the amplitude of the Fourier component of the displacement, γ is the interfacial tension, k_B is the Boltzmann constant, L is the lateral system size, and ξ is the lateral correlation length. Aarts and coworkers [32,33,38] studied the fluid–fluid interface in polymer dispersions by means of confocal laser scanning microscopy (CLSM). They employed poly(methyl)methacrylate (PMMA) and polymeric nanoparticles labeled with light-emitting dyes. Poly(methyl)methacrylate was monitored with confocal laser scanning microscopy and they revealed induced capillary waves at the interface in a phase-separated liquid–liquid interface. Colloidal suspensions have increasing roughness of the interfacial boundary, and polymer addition to a colloidal suspension results in a phase separation between (1) a rich-in-colloid phase and poor-in-polymer phase and (2) a poor-in-colloid phase and rich-in-polymer phase. In the work by Aarts and coworkers, capillary waves at a liquid–gas interface observed through confocal laser scanning microscopy were represented by the intensity distribution of fluorescent light at different time intervals. They also imaged the coalescence of a liquid droplet with the bulk liquid at different time intervals for a total time of 34 s. They demonstrated formation of capillary waves at the liquid–gas interface in a phase-separated colloid–polymer mixture with fluorescent properties. They recorded images through confocal laser scanning microscopy in four states. The viewing plane, perpendicular to the interface, had a thickness of ≈0.6 μm.

Verhoeff and coworkers demonstrated an "optical trapping" of interfaces using polystyrene colloid materials and systems with low interfacial

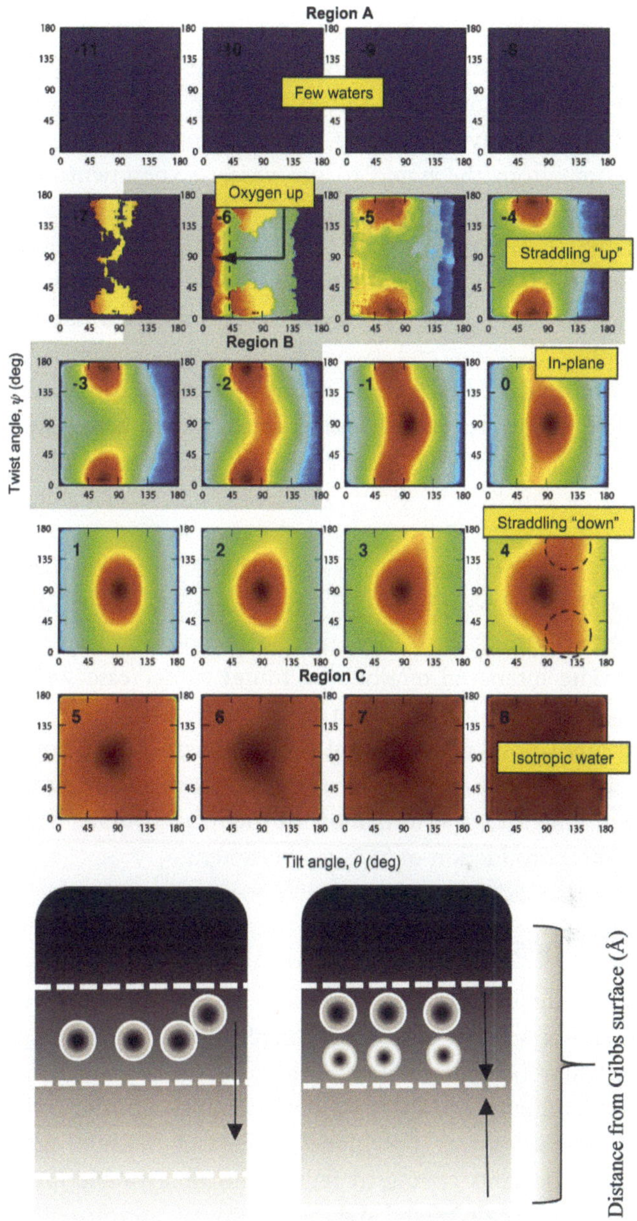

Figure 2.11 A series of snapshots of water molecules at a water–chloroform interface. Observation deep into the organic phase demonstrates that aqueous molecules have their symmetry axes directed toward the bulk water phase. The negative values of the bold numbers [distance (in angstroms) from the Gibbs dividing surface] are representing the bulk organic side and positive the bulk aqueous side. *Reproduced with permission from D.K. Hore, D.S. Walker, L. MacKinnon, G.L. Richmond, Molecular structure of the chloroform – water and dichloromethane – water interfaces, J. Phys. Chem. C 111 (2007) 8832. ©2007, American Chemical Society.*

tension [38]. The gradient forces of a strongly focused $\lambda_{em} = 1064$ nm laser were used to controllably distort and trap the interface. The laser power that was used was 18.5 mW. Deformation took place by moving the optical trap from the liquid to gas phase. Experiments were carried out in heavy water because it does not absorb in the near-infrared wavelengths of the laser tweezer. Small differences in the organic phase may have subtle effects on the structure of the interface and ion transfer through it. Hore and coworkers studied the molecular structure of the interface between water and two closely related solvents: chloroform and dichloromethane [36]. They realized that the orientation of the organic phase and water molecules generates a field across the interface. Compared with the two (rather similar) organic molecules, this field appeared to be much more complicated in the case of dichloromethane (Fig. 2.11).

Another interesting interfacial system is the water–methane system, both in liquid and vapor phase. [37] The system was also studied in terms of molecular dynamics simulations in the presence and absence of SDS surfactants. The interfacial tension was found to decrease with increasing SDS packing fraction for both water–methane and water–vapor interfaces. The simulations demonstrated both the arrangement of the surfactant molecules and the dissolution of the methane in the aqueous phase. It was also found that the formation of methane hydrate was five times faster after the introduction of SDS to the system.

2.6 Conclusions

A series of synthetic pathways can be developed taking into account that the interface is a separate entity of variable thickness, homogeneity and polarity that may be tuned according to addition or in situ formation of nanoparticles and micelles. From various organic molecules available, fluorocarbon materials display the highest interfacial tension against water, with various surfactants aiming to reduce this unfavorable interaction, be it a single surfactant or mixed systems. Study of fluorinated materials as interfacial entities may be important for niche applications involving CO_2–water emulsions due to the unusual and unique solubility of fluorocarbons to supercritical carbon dioxide. Ionic liquids are also interesting systems with versatile structural designs and, in most cases, are immiscible with water, and so form other types of interfaces. Nevertheless there are examples of water-soluble ionic liquids. They have been proposed as alternative solvents with low vapor pressure as well as high viscosity and boiling point.

The fundamental study of these phenomena is an important step for the tailored design of nanostructures, polymeric nanomaterials and self-assembly of particles for formation of thin films in an interface, as well as for unconventional systems where the oil and water transport through minerals is important [39] such as enhanced oil recovery using supercritical fluids in extreme environments. New methods for estimation of interfacial tension as well as for the nature and structure of the two-phase boundary (e.g., micropotentiometry, atomic force microscopy) for systems with very low interfacial tension ($<10\ \mu N/m$) can enable measurement using a very small amount of sample.

The understanding of the interfacial processes can also be providing insights on the nature of water's hydrogen bonding network. This can be done through even-order nonlinear optical spectroscopies, such as vibrational sum frequency generation (VSFG) spectroscopy and heterodyne detected phase-sensitive sum frequency generation (PS-SFG) spectroscopy. This is very important in the field of atmospheric chemistry, the behavior of aqueous aerosols, and the adsorption of ions at the water—minerals interface [40].

References

[1] K. Tauer, S. Kozempel, G. Rother, The interface engine: experimental consequences, J. Colloid Interface Sci. 312 (2007) 432.
[2] K. Bessho, T. Uchida, A. Yamauchi, T. Shioya, N. Teramae, Microenvironments of 8-anilino-1-naphthalenesulfonate at the heptane-water interface: time-resolved total internal reflection fluorescence spectroscopy, Chem. Phys. Lett. 264 (1997) 381.
[3] S. Ishizaka, N. Kitamura, Time-resolved total internal reflection fluorometry study on chemical and structural characteristics at water/oil interfaces, Bull. Chem. Soc. Jpn. 74 (2001) 1983.
[4] H. Wang, E. Borquet, K.B. Eisenthal, Generalized interface polarity scale based on second harmonic spectroscopy, J. Phys. Chem. B 102 (1998) 4927.
[5] L.E. Bakeeva, L.L. Grinius, A.A. Jasaitis, V.V. Kuliene, D.O. Levitsky, E.A. Liberman, et al., Conversion of biomembrane-produced energy into electric form. II. Intact mitochondria, BBA Bioenerg. 216 (1970) 13.
[6] Z. Samec, V. Mareček, J. Weber, Charge transfer between two immiscible electrolyte solutions: Part IV. Electron transfer between hexacyanoferrate(III) in water and ferrocene in nitrobenzene investigated by cyclic voltammetry with four-electrode system, J. Electroanal. Chem. 103 (1979) 11.
[7] C.N.R. Rao, K.P. Kalyanikutty, The liquid—liquid interface as a medium to generate nanocrystalline films of inorganic materials, Acc. Chem. Res. 41 (2008) 489.
[8] E. Rabinowitch, J. Weiss, Reversible oxidation of chlorophyll, Proc. R. Soc. B 124 (1937) 277.

[9] S. Ishizaka, Y. Nishijima, N. Kitamura, Anal. Bioanal. Chem. 386 (2006) 749.

[10] a) P. Shewmon, Diffusion in solids, Minerals, Metals and Materials Society (1989).
b) N.T.K. Thanh, N. Maclean, S. Mahiddine, Mechanisms of nucleation and growth of nanoparticles in solution, Chem. Rev. 114 (2014) 7610.
c) C.D. Crowe, C.D. Keating, Liquid – liquid phase separation in artificial cells, Interface Focus 8 (2018). 20180032.
d) V.P. Shanbhag, C.G. Axelsson, Hydrophobic interaction determined by partition in aqueous twophase systems: partition of proteins in systems containing fatty-acid esters of poly(ethylene glycol), Eur. J. Biochem. 60 (1975) 17–22.
e) D.N. Cacace, A.T. Rowland, J.J. Stapleton, D.C. Dewey, C.D. Keating, Aqueous emulsion droplets stabilized by lipid vesicles as microcompartments for biomimetic mineralization, Langmuir 31 (2015) 11329–11338.

[11] a) J.M. DeSimone, Practical approaches to green solvents, Science 297 (2002) 799.
b) P.S. Shah, J.D. Holmes, R.C. Doty, K.P. Johnston, B.A. Korgel, Steric stabilization of nanocrystals in supercritical CO_2 using fluorinated ligands, J. Am. Chem. Soc. 122 (2000) 4245.

[12] K. Nochi, A. Yamaguchi, T. Hayashita, T. Uchida, N. Teramae, J. Phys. Chem. B 106 (2002) 9906.

[13] a) F. Garbassi, M. Morra, E. Occhiello, Polymer *Surfaces: From Physics to Technology*, John Wiley & Sons, New York, 1994.
b) D.G. Triezenberg, R. Zwanzig, Phys. Rev. Lett 28 (1972) 1183.
c) A.D. Rey, Langmuir 22 (2006) 3491.
d) Y. Sutjiadi-Sia, P. Jaeger, R. Eggers, J. Supercritical Fluids 46 (2008) 272.

[14] S. Fisk, B. Widom, Structure and free energy of the interface between fluid phases in equilibrium near the critical point, J. Chem. Phys. 50 (1969) 3219.

[15] F.M. Fowkes, Determination of interfacial tensions, contact angles, and dispersion forces in surfaces by assuming additivity of intermolecular interactions in surfaces, J. Phys. Chem. 66 (1962) 382.

[16] L. Makkonen and J. Kurkela, *J. Coll. Int. Sci.* **529**, 2018, 243.

[17] B.J. Park, E.M. Furst, Fabrication of unusual asymmetric colloids at an oil – water interface, Langmuir 26 (2010) 10406.

[18] a) S.D. Hudson, J.T. Cabral, W.J. Goodrum, K.L. Beers, E. Amis, Microfluidic interfacial tensiometry, J. Appl. Phys. Lett. 87 (2005) 1.
b) A.G. Banpurkar, K.P. Nichols, F. Mugele, Electrowetting-based microdrop tensiometer, Langmuir 24 (2008) 10549.
c) D.W. Lee, D.J. Im, I.S. Kang, Measurement of the interfacial tension in an ionic liquid–dielectric liquid system using an electrically deformed droplet, J. Phys. Chem. C 117 (2013) 3426.
d) D.L. Williams, C.L. Jupe, K.D. Kuklenz, T.J. Flaherty, An inexpensive, digital instrument for surface tension, interfacial tension, and density determination, Industrial Eng. Chem. Res. 47 (2008) 4286.
e) M. Korenko, F. Šimko, Measurement of interfacial tension in liquid – liquid high-temperature systems, J. Chem. Eng. Data 55 (2010) 4561.

[19] M. Antonietti, S. Forster, Vesicles and liposomes: a self-assembly principle beyond lipids, Adv. Mater. 15 (2003) 1323.

[20] a) E. Stahl, K.W. Quirin, D. Gerard, Dense Gases for Extraction and Refining, Springer Verlag, Berlin, 1987.
b) M. Budich, Countercurrent Extraction of Citrus Aroma from Aqueous and Nonaqueous Solutions Using Supercritical Carbon Dioxide (Ph.D. thesis), Hamburg University of Technology, Germany, 1999.
c) M.F. Kemmere, T. Meyer, Supercritical Carbon Dioxide in Polymer Reaction Engineering, Wiley-VCH Verlag GmbH, Weinheim, 2005.

d) X. Zhang, J.Q. Pham, H.J. Martinez, J. Wolf, P.F. Green, K.P. Johnston, J. Vacuum Sci. Technol. B 21 (2003) 2590.

[21] P. Raveendran, Y. Ikushima, S.L. Wallen, Polar attributes of supercritical carbon dioxide, Acc. Chem. Res. 38 (2005) 478.

[22] K.P. Johnston, K.L. Harrison, M.J. Clarke, S.M. Howdle, M.P. Heitz, F.V. Bright, et al., Water-in-carbon dioxide microemulsions: an environment for hydrophiles including proteins, Science 271 (1996) 624.

[23] a) M.A. Kane, G.A. Baker, S. Pandey, F.V. Bright, Performance of cholesterol oxidase sequestered within reverse micelles formed in supercritical carbon dioxide, Langmuir 16 (2000) 4901.
b) G.B. Jacobson, C.T. Lee, S.R.P. da Rocha, K.P. Johnston, Organic synthesis in water/carbon dioxide emulsions, J. Organ. Chem. 64 (1999) 1207.

[24] M.J. Weida, D.J. Nesbitt, Geometric isomerism in clusters: high resolution infrared spectroscopy of a noncyclic CO_2 trimer, J. Chem. Phys. 105 (1996) 10210.

[25] J.M. DeSimon, E.E. Maury, Y.Z. Menceloglu, J.B. McClain, J.R. Combes, Dispersion polymerizations in supercritical carbon dioxide, Science 265 (1994) 356.

[26] V.H. Dalvi, P.J. Rossky, Molecular origins of fluorocarbon hydrophobicity, Proc. Natl. Acad. Sci. USA 107 (2010) 13603.

[27] a) S. Saraji, L. Goual, M. Piri, H. Plancher, Wettability of supercritical carbon dioxide/water/quartz systems: simultaneous measurement of contact angle and interfacial tension at reservoir conditions, Langmuir 29 (2013) 6856.
b) P.M. Bikkina, Inter. J. Greenhouse Gas Control 5 (2011) 1259.
c) C. Chen, B. Dong, N. Zhang, W. Li, Y. Song, Pressure and temperature dependence of contact angles for co2/water/silica systems predicted by molecular dynamics simulations, Energy Fuels 30 (2016) 5027.

[28] D. Tapriyal, Y. Wang, R.M. Enicka, J.K. Johnson, J. Crosthwait, M.C. Thies, et al., Poly(vinyl acetate), poly(1-O-(vinyloxy) ethyl-2,3,4,6-tetra-O-acetyl-β-d-glucopyranoside) and amorphous poly(lactic acid) are the most CO2-soluble oxygenated hydrocarbon-based polymers, J. Supercrit. Fluids 46 (2008) 252.

[29] a) C. Graham, D.A. Imrie, R.E. Raab, Measurement of the electric quadrupole moments of CO_2, CO, N_2, Cl_2 and BF_3, Mol. Phys. 93 (1998) 49056.
b) P. Bundgen, F. Grein, A.J. Thakkar, Dipole and quadrupole moments of small molecules. An ab initio study using perturbatively corrected, multi-reference, configuration interaction wave functions, J. Mol. Struct. (Theochem.) 334 (1995) 7.

[30] R.S. Hansen, Thermodynamics of interfaces between condensed phases, J. Phys. Chem. 66 (1962) 410.

[31] a) C. Yang, D. Li, A method of determining the thickness of liquid-liquid interfaces, Colloids Surf. A 113 (1996) 51.
b) J.W. Gibbs, The Scientific Papers of J. Willard Gibbs, Volume 1, Dover, New York, 1961.
c) M.K. Berkenbusch, I. Cohen, W.W. Zhang, Liquid interfaces in viscous straining flows: numerical studies of the selective withdrawal transition, J. Fluid Mech. 613 (2008) 171.
d) R. Defay, I. Prigogine, A. Bellemans, D.H. Everett, Surface Tension and Adsorption, Wiley, New York, 1966.
e) S. Marcelja, N. Radic, Repulsion of interfaces due to boundary water, Chem. Phys. Lett. 42 (1976) 129.

[32] M.T. Stone, S.R.P. da Rocha, P.J. Rossky, K.P. Johnston, Molecular differences between hydrocarbon and fluorocarbon surfactants at the CO_2/water interface, J. Phys. Chem. B 107 (2003) 10185.

[33] D.G.A.L. Aarts, M. Schmidt, H.N.W. Lekkerkerker, K.R. Mecke, Microscopy on thermal capillary waves in demixed colloid-polymer systems, Adv. Solid. State Phys. 45 (2005) 15.

[34] D.G.A.L. Aarts, M. Schmidt, H.N.W. Lekkerkerker, Direct visual observation of thermal capillary waves, Science 304 (2004) 847.
[35] M.V. von Smoluchowski, Ann. Phys. 25 (1908) 205.
[36] L. Mandelstam, Ann. Phys. 41 (1913) 609.
[37] D.K. Hore, D.S. Walker, L. MacKinnon, G.L. Richmond, Molecular structure of the chloroform − water and dichloromethane − water interfaces, J. Phys. Chem. C 111 (2007) 8832.
[38] Z.-Y. Lin, D.T. Wu, S.-T. Lin, Equilibrium and transport properties of methane at the methane/water interface with the presence of SDS, J. Phys. Chem. C 122 (2018) 29259.
[39] A.A. Verhoeff, F.A. Lavergne, D. Bartolo, D.G.A.L. Aarts, R.P.A. Dullens, Optical trapping of interfaces at ultra-low interfacial tension, Soft Matter 11 (2015) 3100.
[40] J. Zhao, G. Yao, S.B. Ramisetti, R.B. Hammond, D. Wen, Molecular dynamics investigation of substrate wettability alteration and oil transport in a calcite nanopore, Fuel 239 (2019) 1149.
[41] A.M. Jubb, W. Hua, H.C. Allen, Organization of water and atmospherically relevant ions and solutes: vibrational sum frequency spectroscopy at the vapor/liquid and liquid/solid interfaces, Acc. Chem. Res. 45 (2012) 110.

Further reading

S.G. Booth, R.G. Alghamdi, D. Belić, M. Brust, Electrodeposition of gold nanostructures at the interface of a pickering emulsion, Chem Electro Chem 5 (2018) 2055.
J. Bulou, J.M. Raynal, J. Vial, An interfacial tension treatment with an improved Good—Girifalco equation using solubility parameters, J. Colloid Interface Sci. 98 (1984) 168.
A.J.M. Yang, P.D. Fleming, J.H. Gibbs, Molecular theory of surface tension, J. Chem. Phys. 64 (1976) 3732.

CHAPTER 3

Self-organization of nanomaterials at the interface

3.1 Introduction

After a discussion on the fundamentals that underlie the biological and physical chemistry aspects of interfaces, this chapter presents a thorough analysis on the self-organization processes of nanomaterials and composite layers. The self-assembly of functionalized nanomaterials can be achieved by utilizing immiscible liquids, surfactants, electrochemical redox processes, as well as through manipulation and variation of the ionic strength of these solutions. These processes result in "tailored," bespoke, ordered arrays of nanoparticles (NPs) with unique collective properties. This chapter outlines how the careful selection of organization processes can even, after careful steps, lead to biomimetic and bioinspired functional "nanomachines." This chapter begins with an introduction on the formation of the most fundamental systems, the micelles, and the orientation of surfactants at an interface in the pristine form or when attached to NPs. This chapter continues with the formation of thin films, the utilization of "nanoplasmonics" as building blocks for interfacial films, the deoxyribonucleic acid (DNA)-controlled self-assembly of nanomaterials, and fullerene-based organic–inorganic vesicles.

Furthermore, dual surfactant systems are presented, especially those systems that are composed from different organic molecules, such as fluorocarbons and hydrocarbons. Many of these systems have been studied by the neutron-reflectivity methods to identify their interfacial assemblies. Surfactants can also self-assemble at a dynamic air–water interface, and they can find potential applications in food industries, cosmetics, and insulation materials. In most cases foams are stabilized by proteins or surfactants, and the pH and the temperature affect the interfacial behavior of the system. Recent works have studied the formation of multilayer surfactant structures at the air–water interface using tensiometry and neutron reflectivity, and various surfactants (as dioctyl sulfosuccinate sodium, didecyldimethylammonium bromide, sodium laureth sulfate) presented multilayering phenomenon over a wide range of systems, with and without

Generation of Polymers and Nanomaterials at Liquid-Liquid Interfaces
DOI: https://doi.org/10.1016/B978-0-12-819491-1.00003-4

additives. Petkova et al. [1] found that above a "transitional concentration," foams were found to be stable. This transitional concentration is 10–30 times below the critical micelle concentration (CMC) for sodium dodecyl sulfate (SDS), dodecyltrimethylammonium bromide, and Brij 35. To quantify the properties of the gas–liquid interface, neutron and X-ray reflectivity studies have been performed. By employing small-angle neutron scattering (SANS), both wet and dry foams generated with SDS ($C_{12}SO_4Na$) with different concentrations were studied. Hexadecyltrimethylammonium bromide (CTAB) and polysaccharide/CTAB complex–stabilized free draining foams have also been studied by SANS, and it was shown that the addition of polysaccharide initiates a shift in the peak position toward lower Q values, and the thickness of these liquid films was found to be sensitive to pH and surface charge. The surfactant concentration in these experiments was kept above the CMC, and micelles and foam-induced surfactant lamellae are expected to be present. Foams stabilized by hermo-sensitive fatty acids (12-hydroxy-stearic acid) using SANS and with H_2O/D_2O compositions from both scattering and reflectivity points of view have also been reported as also the interfacial structure of the ethanolamine salt of the same thermo-sensitive fatty acid. The data showed fringes with a layer of thickness of ≈ 300 Å.

The collective properties of NPs in self-assembled arrays are expected to be superior to those of randomly arranged individual particles and could be "tunable" depending on a series of interactions and consequently more diverse compared with their colloidal counterparts. Hence, the assembly of films in interfaces is of special interest. An analysis of the unique "gyroid" structures and their exquisite assemblies as observed in man-made laboratories or biological systems is presented. Finally, the role of two-phase systems in the emerging field of biomimetic engineering is discussed at the end of this chapter.

3.2 The formation of micelles

Before presenting the discussion of more specialized, NP-based and polymer-based systems, I will first provide a comprehensive introduction to the structure of micelles at the water–oil interface. This is the simplest form of a two-phase boundary that is stabilized by an amphiphilic molecule (namely a surfactant). In real-life applications, a more complicated picture emerges when salts, small molecules, or macromolecules are added to this system. In fact, surfactants in an oil–water system may arrange themselves in various forms such as micelles, vesicles, bilayers, or liquid crystalline structures [2a]. The simplest definition of a micelle is the following: an assembly of surfactants having their hydrophobic tails in contact with each other. An inverse micelle (e.g., amphiphilic molecules) has its hydrophilic heads in contact with each other.

For the case of transportation through triple-phase boundaries, see Ref. [2b]. The fundamental notion is that a surfactant with its hydrophobic tail and hydrophilic head lowers the interfacial tension between two immiscible liquids significantly. Various experimental and theoretical methods (e.g., those including Monte Carlo calculations) have been employed to explain the physical observations in these three-phase systems [3]. Fig. 3.1 is a schematic representation of surfactant orientation at the interface. Smit and coworkers [2a] carried out Monte Carlo simulations in a system consisting of 39,000 particles and compared the results with the theoretical data regarding neutron scattering [4]. They demonstrated spontaneous formation of micelles in the aqueous phase and of a surfactant monolayer in the interface. For a density profile of 3%, it was shown through simulations studies that the surfactants are absorbed initially on the interface as monolayers with the hydrophilic head pointing toward the aqueous phase and formation of a depletion layer that contains no surfactants but instead only a small number of water molecules. However, in contrast to the aqueous phase, this depletion layer shows oscillations. This depletion layer between the monolayer and the micelles is a strong indication of short-range repulsion forces between the monolayer in the interface and the micelles in the aqueous phase. These repulsive forces were

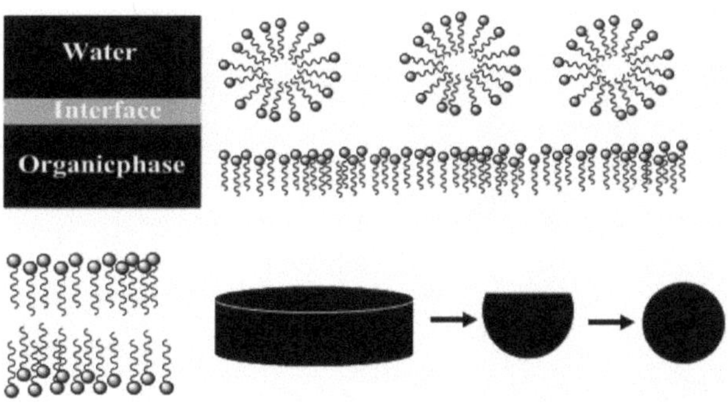

Figure 3.1 (A) The two different arrangements of surfactants in the oil–water system. In the upper aqueous phase, the surfactants form micelles, whereas in the interface, they form layered assemblies. (B) The schematic illustration of the formation of lipid bilayers and closing of their structure to form vesicles.

observed first for lecithin layers with, on average, 21 water molecules per lecithin molecule and which were expanded further to protein–protein interactions within a membrane [5–7]. Here, we have to remind the generic use of the term lecithin to cover any group of plant- or animal-based amphiphilic molecules. The presence of these water molecules around the organic additive was observed experimentally with various spectroscopic and calorimetric methods such as nuclear magnetic resonance spectroscopy and differential scanning calorimetry.

Surfactants may arrange themselves in a different way in a two-phase system even if they are coated and chemically bonded on the surface of NP [8]. Yamamoto and coworkers used total internal reflection surface-enhanced Raman spectroscopy (SERS) to study the orientation of oleate anions capping silver NP, a water–organic phase membrane. The two arrangements are shown in Fig. 3.2. The authors reached this conclusion through comparisons of SERS spectra in the two solutions and the interface, and they monitored the rearrangement and the formation of a "Janus" particle structure.

Surfactants tend to assemble in micelles above the CMC, with hydrocarbon chains directed toward the center of the micelle and the polar heads toward water. The CMCs for four of the most important surfactants are presented in Table 3.1 and Fig. 3.3. The surfactants presented are the organophilic Triton X-100, SDS, and two block copolymers, the pluronic F68 and F127. "Pluronics" (also known as "polyoxamers") are nonionic triblock copolymers with a central hydrophobic chain (polypropylene oxide) and two side chains of hydrophilic polyethylene oxide. They were invented by Irving Schmolka in 1973.

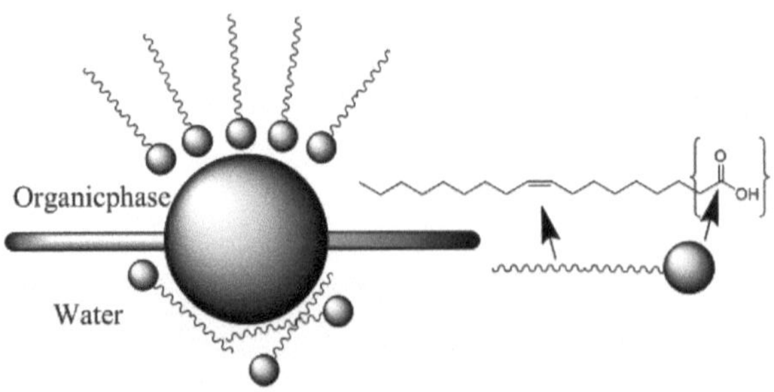

Figure 3.2 The different arrangement of organic surfactants, for example, oleic acid, on the surface of silver NP in the interface. In the aqueous phase, the double bond is attached on the particles, whereas in the organic phase, the carboxylate anion and hydrocarbon chain are directed toward the organic solvent [8].

Table 3.1 The critical micelle concentration in mM or in mg/mL of four of the most commonly employed surfactants.

Surfactant	CMC (mM)
Triton X-100	0.22 mM (0.162 mg/mL)
SDS	8.2 mM (2.36 mg/mL)
Pluronic F68	100 mg/mL
Pluronic F127	25 mg/mL

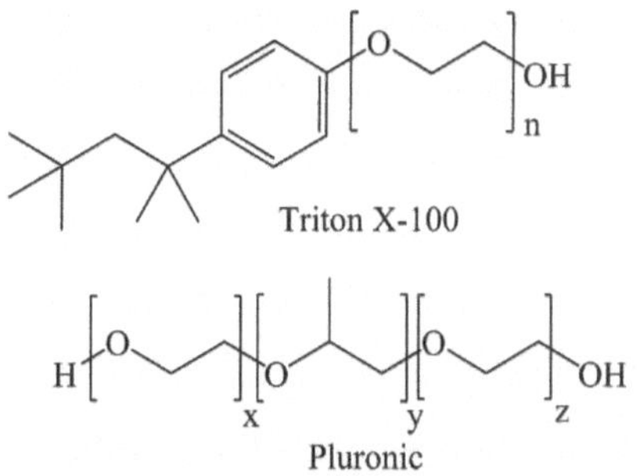

Figure 3.3 The CMC (in mM or in mg/mL) of four of the most commonly employed surfactants, Triton X-100, SDS and the two pluronic block copolymers, and the structure of Triton X-100 and the general block copolymer structure for the family of pluronics.

3.2.1 Catanionics: mixture of anionic and cationic surfactants

The neutral and the ionic surfactants have been investigated thoroughly as micellar assemblies and counter ions. In addition, the considerable activity has been dedicated to mixtures of surfactants. The mixture of different types of surfactants is exciting for fundamental studies, field tests, and industrial applications. An interesting combination would be a hydrocarbon and a fluorocarbon surfactant due to the hydrophobicity of both components and the different packing arrangement that they would inherit at an interface or in a one-phase solution. Simister and coworkers compared, through a series of surface tension and neutron scattering methods, the adsorption of ammonium perfluorooctanoate (APFO) and ammonium decanoate (AMDEC) at the air–solution interface (Fig. 3.4) [9a]. Their methodology was based on the refraction of neutrons if there is a change in the refractive index between the two media as they travel from one to the other. They calculated the areas occupied by the surfactants at a pH of 8.8, and the ionic strength of the NH_4Cl-NH_4OH was 0.1. They found that these areas were 41 Å^2 for APFO and 35 Å^2 for

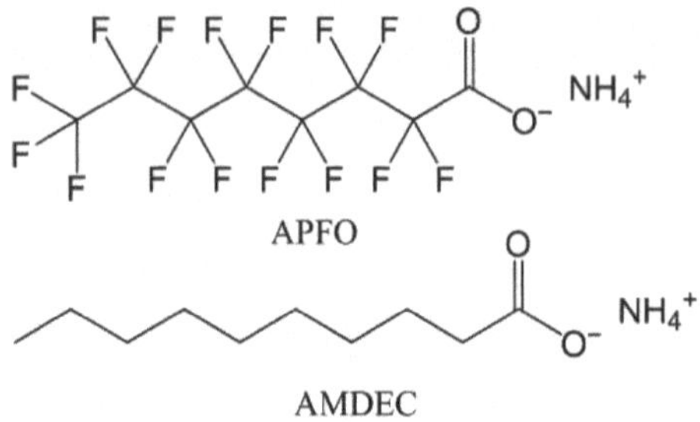

Figure 3.4 Structure of a perfluorinated and hydrocarbon ammonium surfactants. For details on these systems, see the work by Simister and coworkers [9].

AMDEC. Furthermore, their studies demonstrated that the APFO micelles have a different arrangement, forming cylinders instead of spheres. Their molecular structure, density, and neutron scattering length densities are shown in Fig. 3.2. The area per molecule at the surface is calculated using Eq. (3.1):

$$A = \sum nibi/\rho s\tau, \tag{3.1}$$

where n_i is the number of atoms at the scattering length b_i, and ρ_s is the density of the layer. In conclusion, neutron reflectivity studies provide a good overall picture for absorption at the air–surface layer and arrangement of the molecules there.

A mixture of anionic and cationic surfactants is termed "catanionics." This mixture possesses a different behavior and physicochemical properties compared with the neat counterparts [9–13]. These cosurfactant mixtures lead to the formation of spherical, rod-like, elongated, micelles and vesicle microparticles. These mixtures above the CMC are good laboratory-based examples of cell membranes. The size and shape of the particles depend on the concentration and the type of the two surfactants, with vesicles formed at 1–2 wt.% due to favorable thermodynamics and ordering parameters [14–16].

3.3 Organic–inorganic vesicles

Liposomes derive their name from two Greek words, λίπος (meaning "fat") and σώμα (meaning "body"). The term "vesicles" is derived from the Latin word, meaning "small bubbles." For most of the people, the two words have the same meaning [17a]. Phospholipids can arrange themselves in vesicles and lamellar morphologies based on the hydrophobic effect and concentration [17b]. Block copolymers have been proposed as a system that can mimic the arrangements of phospholipids. For example, the block copolymer $E_{115}B_{103}$ ("E" is the oxyethylene unit, OCH_2CH_2-, and "B" is an oxybutylene unit, $-OCH_2CH$ $(CH_2CH_3)-$, and the subscripts are the number average degree of polymerization) can form stable vesicles with a membrane thickness of 7.5 nm. The same block copolymer forms a hexagonal melt at a temperature of 70°C with a spacing of 17.7 nm. For a detailed analysis, refer the work by Battaglia and coworkers [17b]. There are block copolymers with various number average degrees of polymerization, each one of them able to self-assemble in different structures. Detailed studies on E/B diblock copolymers and their crystallization behavior have been published by Ryan and coworkers, who derived their conclusions through simultaneous small-angle X-ray scattering (SAXS) and wide-angle X-ray scattering [17c]. The first work on block copolymer vesicles was developed by the group of Aizenberg, who is better known for the impressive biomineralization studies. They employed a polymer as a tectonic block with low glass transition temperature (T_g) to achieve the ordered state.

Vesicle formation using exotic carbon nanomaterials has been presented later in this chapter. Fullerenes are fascinating small molecules with high symmetry and exciting optical, chemical, and electrochemical properties. In Chapter 1, Hydrophilic and Organophilic Systems in Nature, I outlined their unconventional synthesis and their unexpected encounters in interstellar media. I will start by outlining the formation of fullerene vesicles in two-phase media with functionalized "buckyballs" as reported by the group of Nakamura [18]. The polar–nonpolar head – tail system is the fundamental biological unit. Artificial lipid vesicles can be of great importance in biomedical applications for controlled and light-triggered release of drugs and site-specific cell imaging. Nakamura and coworkers synthesized a series of novel nonpolar/polar/nonpolar hierarchical structures. In their work, 20 potassium complexes were tested. The complexes were based on potassium and penta-[(4-substituted) phenyl][60]fullerene anions that acted as potential bilayer vesicles. They were functionalized with alkyl groups: methyl to icosanyl, perfluoromethyl, perfluorobutyl, and perfluorooctyl. They consist of a bilayer of amphiphilic molecules. They observed that, despite the highly hydrophobic nature of fullerene and alkyl/perfluoroalkyl groups, all compounds except for the one bearing perfluoromethyl groups appeared to be soluble in water. This was with regard to the cyclopentadienide part of the system. The systems formed a NP vesicle with a diameter of 25–60 nm. Fig. 3.5 shows the structure of the C_{60}-based vesicles. They were studied with dynamic light scattering, atomic force microscopy (AFM), and scanning electron microscopy.

Further expanding their work, the authors used fullerene-based vesicles as scaffolds for self-assembly of gold plasmonic NP. They employed vesicles with a diameter of 30 nm and covered them with gold particles with a diameter of 3.5–7.2 nm (Fig. 3.6). Colloidal silver NP and silver–polymer nanocomposites are utilized as antibacterial agents. However, a gold NP with surface plasmon resonance could be appealing for the synthesis of mesoscopic assemblies that may show collective properties through coupling of the surface plasmon resonance of neighboring particles. Previous attempts at the synthesis of lipid vesicles with gold particles led to problems

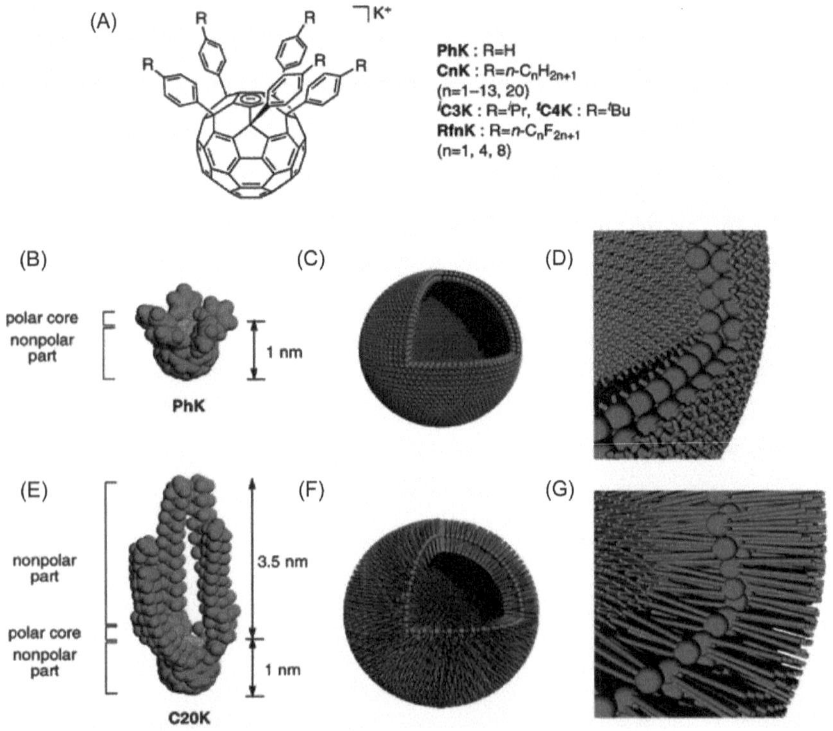

Figure 3.5 (A) Fullerene vesicles (schematic). They comprise C_{60} fullerene decorated with aromatic and aliphatic groups. (B–E) The length of each group and bilayer arrangement. (F) A schematic model of a vesicle. (G) A magnified image of a bilayer of the vesicle. *Reproduced with permission from T. Homma, K. Harano, H. Isobe, E. Nakamura, J. Am. Chem. Soc. 133 (2011) 6364. ©2011, American Chemical Society [18].*

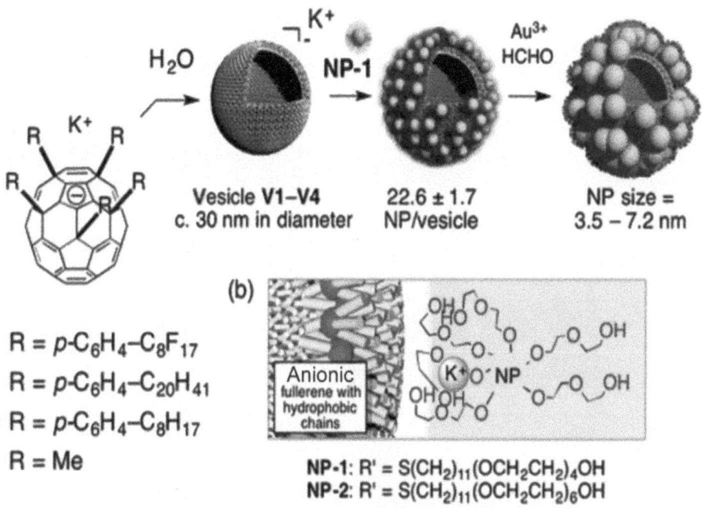

Figure 3.6 A schematic representation of the self-assembly of gold NP on fullerene vesicles. The R represents the four functional groups that could be attached on the sp2 hybridized C_{60} surface. *Reproduced with permission from R.M. Gorgoll, T. Tsubota, K. Harano, E. Nakamura, Cooperative self-assembly of gold nanoparticles on the hydrophobic surface of vesicles in water, J. Am. Chem. Soc. 137 (2015) 7568. ©2015, American Chemical Society [20].*

on the stability and the control of the density and distribution of NP. Other important drawbacks were fusion and aggregation upon decoration with particles. To address these problems, the fullerene amphiphiles $R_5C_{60}-K^+$ were employed (Fig. 3.6). Finally, in a recent study, Bairi and coworkers used the same fullerene vesicles in an ultrasonic mediated liquid – liquid interfacial precipitation for the synthesis of anisotropic nanostructures [18b]. They found that the growth of these tube-like anisotropic structures arises from a combination of interfacial assembly and phase separation.

This is a versatile and highly promising method because these multicomponent and multifunctional NPs are stable up to 90°C in ambient conditions or in vacuum. The small gold NPs, when mixed with vesicles, lead to an increase in the hydrodynamic diameter from 30.4 to 49.4 nm and to a decrease of the ζ potential from −45.6 to −13.7 mV. NP in a droplet can move from the vesicle to the liquid surface, and hence, a decrease in the surface tension from 72 to 50 mN/m was observed. See the work by Gorgoll and coworkers in Figs. 3.6 and 3.7 [19].

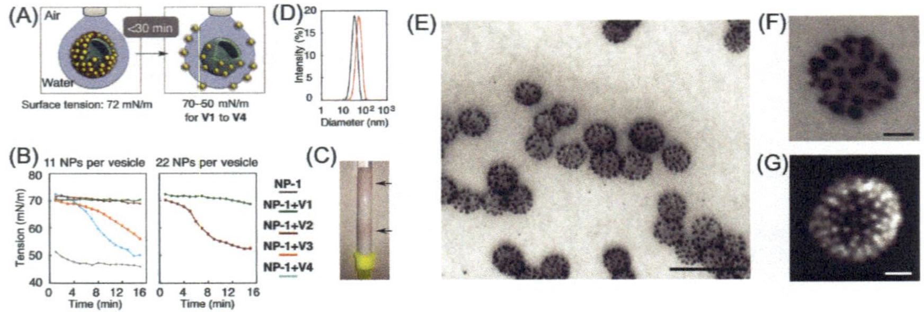

Figure 3.7 Adsorption of gold NP capped with (11-mercaptoundecyl)tetra(ethylene glycol) on fullerene vesicles with four functional groups (1 – 4). (A) There is a reduction in the air–water surface tension caused by the release of gold NP with time according to the transmission electron micrographs and scanning transmission electron micrographs of vesicles. (B) The time dependence of the interfacial tension. (C) Separation of nanoparticles with column chromatography. (D) Dynamic light scattering measurements of the vesicles. (E) STEM image of the vesicles, scale bar: 100 nm. (F and G) STEM and SEM images, scale bar: 10 nm. *Reproduced with permission from R.M. Gorgoll, T. Tsubota, K. Harano, E. Nakamura, Cooperative self-assembly of gold nanoparticles on the hydrophobic surface of vesicles in water. J. Am. Chem. Soc. 137 (2015) 7568. ©2015, American Chemical Society [20].*

A two-dimensional polymerization at an air–water interface provided a direct spectroscopic evidence regarding the type of the cross-links that lead to the large-scale formation of a covalent framework nanolayer. Various techniques have been used for the study of this framework. Examples include the tip-enhanced Raman spectroscopy and mapping after a transfer of both the monomer and the polymer monolayer onto Au (111). This transfer and the subsequent study allowed for an estimation of the crystallinity through means of percolation theory and by locating the regions that exhibit defects [20a].

The gel trapping technique (GTT) has been used for the identification and calculation of the three-phase contact angles of silica particles in nanoscale dimensions that were partially decorated with organophilic octadecyl groups. Contact angles of colloid silica and gold particles at air–water and oil–water interfaces have also been determined with the GTT. These silica particles were self-assembled at air–water and decane–water surfaces. Then, the aqueous phase was subsequently gelled with a nonadsorbing polysaccharide and were lifted by molding with curable poly(dimethylsiloxane). SEM was employed for the determination of the particle contact line diameter. Furthermore, the authors used the same techniques to study the contact angles of agglomerated gold powder microparticles at the air–water and the decane–water interfaces, and they found that these NPs demonstrate considerably higher contact angles than those that were assembled and dried on flat gold-coated surfaces [20b]. Shehri et al. have studied the absorption of carboxylic modified latex microparticles at liquid–liquid surfaces. They identified the preferred type of emulsion that can be efficiently stabilized. This was found to depend on the particle size and the surface density of the carboxylic groups that were decorating the latex particle. Similar to the work by Cayre et al., they measured the particle contact angle by using the GTT for the latex particles that were adsorbed at air–water and oil–water interfaces, and they subsequently recorded the SEM images of the polydimethylsiloxane replicas of the liquid interface with the particles. The Pickering emulsions were found to be more stable with the increased density of the carboxylic acid groups [20c].

3.4 Deoxyribonucleic acid–directed self-assembly

DNA macromolecules provide a platform with unique functionalities and self-assembly opportunities. The exquisite DNA *origami* was presented by

Rothermund [21]. This nanoscale folding of DNA through selective interactions between base pair has subsequently influenced materials science, biology, and art.

The organization and folding of DNA may also be utilized for the self-assembly of nanomaterials at an interface. Srivastava and coworkers directed the assembly of NP at an interface driven by interaction among DNA macromolecules [22,23]. The authors controlled the ionic strength of the solution using sodium chloride as the electrolyte. Through stepwise addition of the salt, they dictated the assembly of NP and the rheological properties of the membrane formed on the interface. When the salt content was low, the loss modulus (G') of the membranes appeared to be higher than the storage modulus (G''), whereas, for high salt content, the opposite was observed. Essentially, they created programmable NP membranes at liquid—liquid interfaces.

The ionic strength has been shown to direct the self-assembly of NP as well as the size and shape of poly(o-phenylenediamine) nanostructures [24]. The behavior of the system is dictated by competition between a charge-induced repulsion and a DNA-induced attraction, and in situ SAXS studies of the interface have revealed various phase transitions. In other examples of self-assembled structures, Sawall and coworkers looked at the arrangement of gold NP, with each cluster consisting of 13 NP [25].

Another example of control of the organization of NP through the ionic strength of the solution is the formation of two-dimensional (2D) Gibbs' monolayers of thiolated single-stranded DNA-functionalized gold NP at a vapor—solution interface. These assemblies were controlled through concentration gradients of magnesium chloride and calcium chloride and were demonstrated through X-ray scattering and X-ray reflectivity studies [26].

3.5 Interface energy and interfacial tension in thin films

In colloid and interface science, particles can decrease the interfacial tension and can form various self-assembled structures in the interface [27]. Due to a reduction of the organic—water interface area, the assembly of NP leads to reduction in the free energy of the system [28]. According to the pioneering work of Whitesides and Boncheva[29], a self-assembling system needs a sufficient degree of reversibility to form highly ordered arrays. If this requirement is not satisfied, then disordered (glassy) assemblies are produced.

The first analysis of the reduction in the interface energy was presented in 1980 [30] by Pieranski. He distinguished between the surface tension among particles ($\gamma_{s/o}$ and $\gamma_{s/w}$) and the organic phase and among particles and the aqueous phase. This effect seems to be dependent on the shape of the particles because, according to studies on interfacial systems, the reduction in the interfacial energy is higher for cylindrical particles compared with spherical particles [31].

Very interesting multicomponent and functional systems are the "Janus" nanomaterials, polymers, and molecules. They take their name from the two-faced Roman God Janus (Fig. 3.8). Some small Janus molecules demonstrate the interesting photophysical properties, such as two-photon absorbance [33]. With regard to their noncentrocymmetric architecture, Janus particles provide unique characteristics such as amphiphilic materials and surfactants for self-organization at the nanoscale. For example, in a work by Ruhland and coworkers, it was demonstrated that the shape of two polystyrene (PS)-*block*-polybutadiene (PBD)-*block*-polymethyl methacrylate (PMMA) block terpolymers and the PS-*block*-PBD-*block*-poly(*tert*-butyl acrylate) block terpolymer influenced the interfacial properties of toluene—water interfaces [34]. They also found that the interfacial tension of the toluene—water interface was dependent on the size, shape, concentration, and packing of Janus particles. Their findings were interesting and could provide aid for the development of other similar systems with different properties. They observed that addition of a cylindrical-shaped Janus particle at the interface led to the maximum reduction in the interfacial tension (from 34 to 14 mN/m). When Janus spheres were added, there was a decrease until 17.5 mN/m, whereas with Janus discs, a decrease occurred until 19 mN/m. In the case of Janus discs, the decrease was rapid initially, but reached a *plateau* at some point before the further decrease. They assigned this unexpected behavior to packing of the discs on the interface. Another factor for this kinetic effect could be the particle size. A systematic study based on the size effect was done, with diameters of spheres being 50 nm, cylinders 2300 nm, and discs 300 nm.

Various architectures were constructed based on these copolymer microparticles. The latter showed a more pronounced decrease in the interfacial energy compared with their conventional counterparts, with cylinders being able to reduce the interfacial tension more than discs or spheres. Sheng and coworkers synthesized a polyethylene glycol-*block*-poly(L-lysine)-*block*-PS triblock copolymer through a combination of

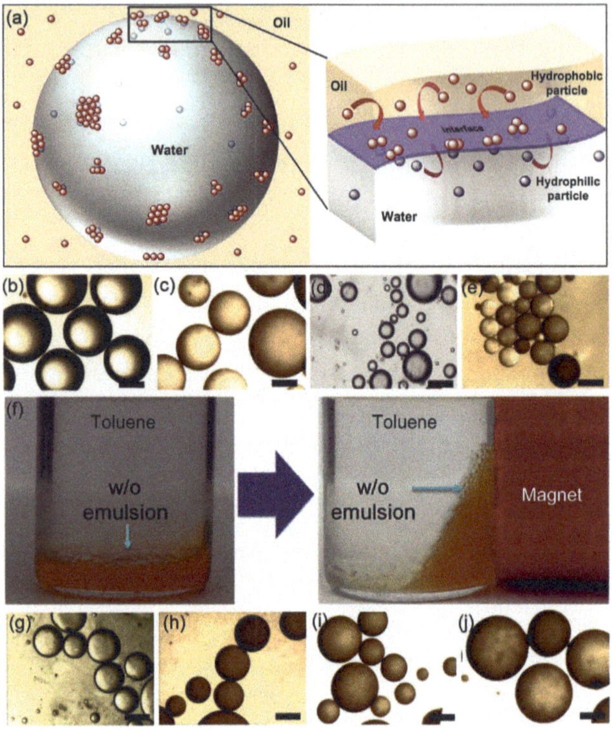

Figure 3.8 (A) Emulsion stabilization of hydrophilic and hydrophobic particles. (B–E) water–in–oil (W/O) emulsion or O/W emulsions forming Janus-like globules. (F–J) the emulsion can be attracted by a magnet. *Reproduced with permission from A. Saha, V.T. John, A. Bose, In situ assembly of hydrophilic and hydrophobic nanoparticles at oil–water interfaces as a versatile strategy to form stable emulsions. ACS Appl. Mater. Interfaces 7 (2015) 21010. ©2015, American Chemical Society [32].*

ring-opening polymerization and atom transfer radical polymerization. Also, they employed this triblock copolymer as a surfactant in oil-in-water emulsions. Subsequently, they prepared Janus silica hollow spheres. Their silica synthesis followed a biomimetic approach with precursors of silicic acid [35a].

Through the pendant drop method, time-dependent studies revealed a substantial decrease of the oil–water interface upon absorption of Janus NP, with larger particles leading to a larger decrease in the interfacial tension [35b]. For example, on addition of a block terpolymer to a cyclohexane solution on a cyclohexane–water system, the interfacial tension between the organic phase and aqueous phase decreased from 51 to ≈ 30 mN/m. Also, with the use of cryo-transmission electron microscopy (which was applied for the first time in tetrahydrofuran), the aggregation behavior of Janus polymer particles was visualized in situ [32a].

Smaller spherical particles showed a higher diffusion coefficient due to the Brownian motion and, consequently, faster absorption kinetics. All particle types showed irreversible adsorption. Discs had two orientations with respect to the water–toluene interface, but Janus cylinders assembled with the PS part pointing toward the organic phase and the PMMA part toward the aqueous phase. The study by Glaser and coworkers on NP capped with dodecanothiol and octadecanothiol demonstrated that the more hydrophobic iron oxide Janus particles elicited a more significant decrease in the interfacial tension [36].

In general, the interfacial energy depends on the repulsion between the NP and each of the two phases as well as on the radius "r" and vertical distance between the center of the particle and the interface. This can be described using Eqs. (3.2)–(3.4):

$$\gamma s/w = \gamma w 2\pi r 2\left[1 + \left(z/r\right)\right] \tag{3.2}$$

$$\gamma s/o = \gamma o 2\pi r 2\left[1 - \left(z/r\right)\right] \tag{3.3}$$

$$EW/o = -\gamma W/o 2\pi r 2\left[1 - \left(z/r\right)2\right] \tag{3.4}$$

Consequently, the reduction in free energy of the interface depends strongly on the particle size. For example, in microparticles, the reduction is 10^7 $k_B \times T$ and in NP is 10^1–10^4 $k_B \times T$. Hence, very small particles tend to move away from the interface, whereas larger particles have a stronger tendency to form an interfacial film. The particle size influences the stability of the system, but the latter also depends on the position of particles at the interface, with maximum stability achieved for $z = 0$ ($\theta = 90$ degrees) [Eq. (3.2)]. A suitable surface modification of the particles with organophilic or hydrophilic ligands may lead to further decrease in the interfacial tension. Thus, the concept developed by Mohwald and coworkers [37] with plasmonic gold, silver, or magnetic g-Fe_2O_3 NP surface functionalized with the ligand 2-bromo-2-methylpropionate resulted in a contact angle of these particles of 90 degrees and assembly of particles exactly in the middle of a toluene–water interface. This thin film was transferable through Langmuir–Blodgett methods and exhibited a metallic gold reflectance and blue transmittance. Furthermore, heterodimeric particles of iron oxide–silver were synthesized at an interface through a controlled reduction in a dichloromethane—water droplet system, and the particle size was 8 and 5.5 nm, respectively [38].

Petkov and coworkers synthesized NP monolayers by exploiting the electrostatic repulsion in air—liquid interfaces. The effect of the surface pressure, Π, on aggregation was associated with the interparticle distance L, as $\Pi \sim L^3$ [38a]. Kralchevsky and coworkers summarized their findings in a second study and developed theoretical models on an asymptotic law that correlates pressure with NP aggregation [39b]. Other works on liquid—liquid interfaces have reported deposition of biodegradable polymer capsules with potential biomedical applications [39c].

Anisotropic nanomaterials may also be synthesized in interfaces. A versatile and very efficient method for creating asymmetrical particles with flattened, flake-like morphology was developed by Park and Furst. The materials were derived from colloidal latex microspheres and an oil—water interface [36]. The authors controlled the shape and the degree of asymmetry and anisotropy through treatment at different temperatures. They provided evidence that the primary force driving the transformation in particle shape is the strong interfacial energy of the oil—water interface. Particle absorption at the interface is irreversible, and subsequently particles are heated to near the T_g, a process that deforms and reduces the contact between the oil phase and water phase. The trapping energy for the particles is high ($\Delta E \approx 10^8\ k_B \times T$) and reflects the large energetic penalty for contact between the oil phase and water phase relative to the particle. Also, the interfacial tension at the organic—aqueous phase interface increases weakly as the temperature decreases (e.g., $\gamma_{ow} = 41$ mN/m at 90°C and 46 mN/m at 60°C) and is independent of the gellan concentration.

Optical properties of NP are influenced by the distance and type of interaction (i.e., head—head, side—side) for nanorods. Hence, the idea of a "plasmonic ruler" was introduced to provide spatial information on NP 2D arrays, dimers, and tethered particles in interfaces [40,41a]. Conversely, three-dimensional (3D) ordered gold NP arrays can exhibit even more exciting medium properties. For example, negative permittivity can be achieved, which leads to metallic optical response with metal content <1% [41b].

The behavior of charge-stabilized NP in solution is dictated by the Derjaguin—Landau—Verwey—Overbeek (DLVO) theory, which treats the electrostatic and van der Waals forces governing these colloidal systems. These interactions, in contrast with interface interactions, are weak and short range [41c]. The DLVO theory explains the stability, aggregation, and subsequent precipitation of aqueous solutions quantitatively by combing the van der Waals attraction mentioned earlier and electrostatic repulsion forces with respect to the double layer of counter ions, whereas mean-field approximations treat the electrostatic part of the interaction with respect to the thermal energy scale. This is particularly important for exploration of underground water. In these underground environments, the water is of particularly high salinity with the increased contents of calcium and magnesium cations as well as phosphate, chlorine, and other anions.

However, a growing interest in colloidal stability and behavior has revealed that the DLVO theory cannot explain the stability of nanomaterials migrating and assembling at the liquid–liquid interface. Interactions between ions, molecules, and particles at the interface are far more complex than those of a homogeneous solution and are governed by interfacial tension, dipole–dipole interactions [42], asymmetric counter ion cloud hydration forces, capillary forces, and thermal fluctuations [43]. Fig. 3.9 presents an overview of the forces acting on the interface and dictating particle movements and film formation.

Emulsions stabilized by NP are known as "Pickering emulsions" and have been the focus of attention in colloidal and interface science for more than one century. Rheological sheer thinning or sheer-thinning behavior has been observed in Pickering emulsions using a double-wall ring interfacial rheometer [45a]. Even cyclodextrins may be employed for stabilized pickering emulsions, as shown in the recent work by Wu and coworkers, who mixed cyclodextrins with 12 types of oils [45b]. Nanofibers from cellulose have been used for the pickering stabilization of biocompatible foams and emulsions even if the fundamental mechanism behind this stabilization is not yet elucidated. Bertsch et al. synthesized Cellulose nanofibrils (CNFs) with varying contour lengths and charge densities, and they studied their absorption and stabilization at the air–-water interface. It was found that these CNFs reduce surface tension and form interfacial layers with high viscoelasticity [45c].

Turek and coworkers utilized a water–1,2-dichloroethane interface and controlled the interparticle distance of gold NP at the interface through centrifugal forces at a distance of 6–35 nm. Also, centrifugation time and electrolyte concentration were two of the parameters considered. For example, NaCl concentration at 1–75 mM gave rise to distinctive surface plasmon visible light absorbance spectra. Finally, to expand

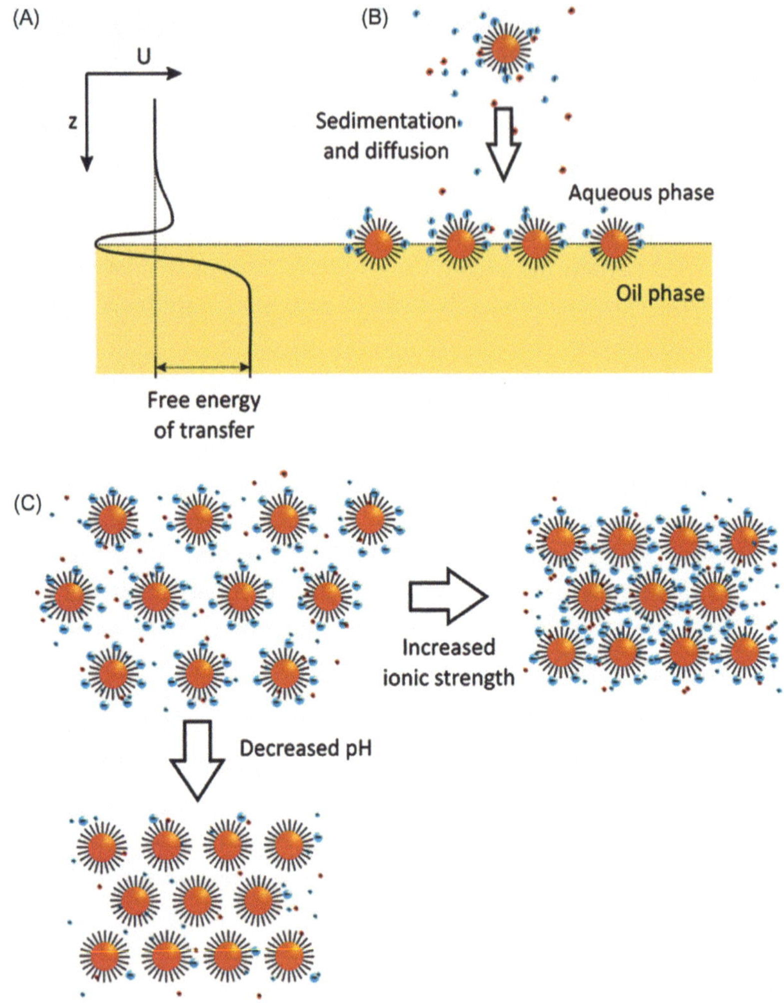

Figure 3.9 Brief representation of the localization of NP functionalized by charged ligands at the liquid–liquid boundary. (A) The free-energy profile, "U" ("z") is dominated by interfacial stabilization. This is due to reduction of the free energy in the system because particles are entering an energetically unfavorable interface ("pickering emulsions"). (B) The interfacial energy traps particles at the interface. (C) Ionic strength controls interparticle separation. *Reproduced with permission from V.A. Turek, M.P. Cecchini, J. Paget, A.R. Kucernak, A.A. Kornyshev, J.B. Edel. Plasmonic ruler at the liquid–liquid interface, ACS Nano 6 (2012) 7789. ©2012, American Chemical Society [44].*

further the concept of self-assembled hybrid nanomaterials, Pothorszky and coworkers attached gold spheres on the tails of gold nanorods and observed substantial change in angle-dependent extinction spectra [46].

Another example of self-assembly directed through the interfacial tension gradient is the fabrication of large-area, ordered assemblies of polymer-capped NP. For example, Che and coworkers, through variation in the concentration (10, 20, and 70 nM) of PS-coated gold NP of diameter 9 nm, studied the assembly of the particles on silicone that were surface modified and had a surface tension between $\gamma s \approx 20$ and ≈ 80 mN/m. Formation of the thin film was also dependent on the concentration and the deposition velocity with the substrate, with the highest surface energy of 80 mN/m leading to the synthesis of monolayers with the highest order [47].

Other interesting systems are the "metastable" assemblies of NP at interfaces. Xue and coworkers created 2D aggregates far from equilibrium [48a] and Sabapathy and coworkers [48b] created from the equilibrium position of NP on interfaces. The assembly of materials in interfaces, such as the heptane–water interface, has been monitored with AFM. Costa and coworkers characterized, using AFM, silica NP with the average diameter of 10 nm [48c]. Particles were coated with cetyltrimethylammonium bromide to obtain an amphiphilic character. Another method used to image these interactions is Brewster angle microscopy, whereas AFM is, in general, a good tool to observe membrane functionalities [48d]. The water–mineral interfaces and the role that ions play on this hydrogen bonded boundary have been studied as well. The effect of the local electric field according to the Chapman model, the pH, and the nature of the ions have been shown to influence the organization of the water and mineral molecules at the interface [48e].

3.6 Self-assembled thin films with plasmonic nanoparticles

Interfaces have been shown to be unique media for the fabrication of membranes and thin films. They enable the formation of thin films based on the particle–particle and particle–surface interactions. Membranes consisting of NP have very good mechanical properties, with a Young's modulus of 1–10 GPa [49]. Their mechanical properties depend on the ligand chains on the surface of the NP. Salerno and coworkers [49c] undertook atomistic molecular dynamics simulations to evaluate the effect of the end groups of the ligands on the mechanical properties of NP membranes. They used gold NP, and for dodecanethiol ligand chains of gold cores (diameter, 6 nm), the Young's modulus was 2.5 GPa. When the CH_3-terminated chains were used, the Young's modulus increased by 50% when the end groups were replaced by COOH. The authors treated the COOH groups as dipoles with a binding energy of 60–80 kJ/mol.

Plasmonic NP, like gold and silver, can be synthesized in various sizes and shapes, thereby giving rise to unique absorption and SERS because of oscillation of the free plasmon electrons on their surface. With respect to their plasmon properties, noble metal NP may behave as "antennae" for light. These localized, confined plasmon modes are excited by incident light, and they fluctuate like dipoles. A single NP will emit light in all directions, but an ordered, hierarchical array of NP generates emitted light that is coherent from NP to NP, thereby enhancing the emission along a specific direction. The first application of the surface plasmon resonance of gold and silver NP dates back to the famous Lycurgus cup made in Rome during the 4th century CE. The cup depicts the death of Lycurgus from Dionysus, and to achieve the dichroic effect, and its creators dispersed gold and silver alloys within the glass matrix. The color of the Lycurgus cup (which can be seen in the British Museum) is different in reflected and transmitted light due to the dispersed gold and silver alloys within the glass matrix.

Jung et al. described a route for the formation of microscale droplets that are covered with amphiphilic gold NPs. The gold NPs that were used for this study were surface functionalized with a mixture of hydrophilic and hydrophobic ligands. First, the amphiphilic gold NPs in the aqueous solution were mixed with hexane. After the mixture was left to settle, the gold NPs migrated to the hexane–water interface. It was found that the molar ratio between the hydrophobic and the hydrophilic surfactants dictates the migration and assembly of the NPs at the interface, following the Langmuir isotherm model [49d].

From the 1960s to early 1980s, the surface plasmon properties (Fig. 3.10) of these colloids were focused on gold NP. They found applications as sensors and SERS agents after the work pioneered by Mirkin. These applications defined the newly developed field of "nanoplasmonics" for biomedical applications (e.g., DNA sensing).

Fig. 3.10 shows the oscillation of the electron cloud on spherical gold NP. The most common water-soluble particles, the citrate-coated gold nanospheres are expected to possess a strong absorbance at 530 nm with a red color that may be tuned upon exchange of capping agents.

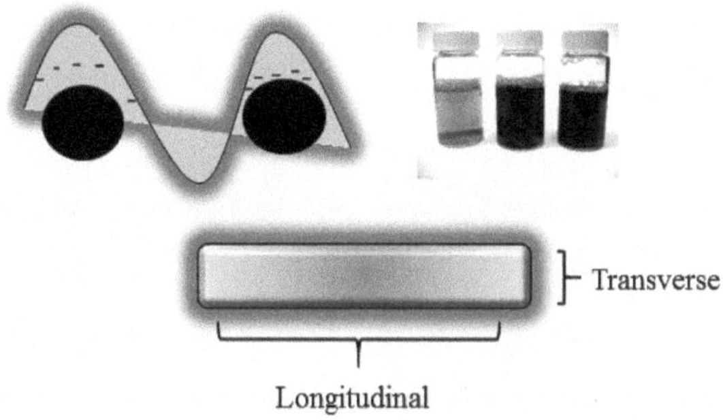

Figure 3.10 The oscillation of the surface electrons on spherical nanoparticles. This graph represents the electron oscillations of the surface plasmon resonance. This effect results in different-colored colloidal solutions depending on the aspect ratio in the case of nanorods. This anisotropic shape gives rise to the longitudinal and transverse oscillation bands.

The electrodynamic theory behind the optical properties of plasmonics can be found in Mie's theory on scattering from spherical-shaped dielectric materials [50]. Fig. 3.10 shows the transverse and longitudinal electron oscillations of gold nanorods.

One interesting application of interfaces in plasmonics is the formation of mirrors consisting of plasmonic arrays. The first report on the formation of silver NP layers on an interface was reported in the early and important works of Yogev and Efrima [51]. They synthesized metal liquid-like films with silver NP having a film thickness of several micrometers at the water–organic interface. Other research team synthesized similar films using water–air interfaces with similar methods [52,53].

This method resulted in a novel class of "liquid mirrors" consisting of noble metal plasmonics. A quasi–D array of NP (diameter, 50 nm) in an interface formed a mirror surface with substantial reflection. The authors studied the dependence of mirror formation on the concentration, size, and shape of NP. This approach is very interesting because, according to classical electrodynamics, a thin layer (≈ 50 nm) of metal cannot be used as a mirror as it will transmit incident light. However, it appears that an inhomogeneous film that composed of isolated NP behaves in a different way.

With regard to the refraction or reflection of light in a two-phase system and an interface, one must take into account that the refractive index of the aqueous phase is affected slightly by the concentration of NP and other materials in it (Fig. 3.11). However, the concentration is low, so

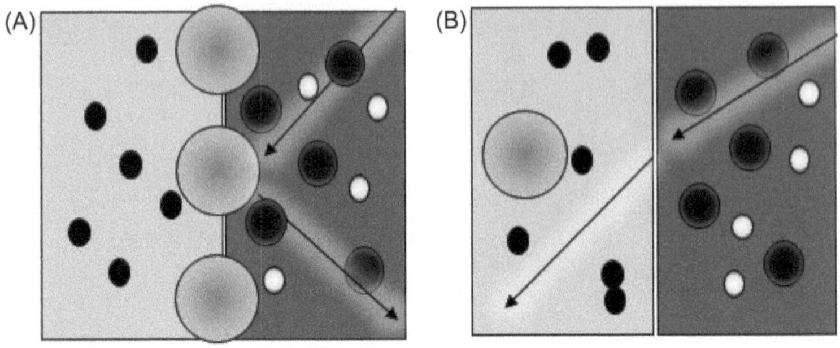

Figure 3.11 (A) A schematic representation of the light reflection that originates from a monolayer consisting of nanoparticles. The beam of incident light is penetrating the system from the oil side (right). The cations and anions of inorganic and organic electrolytes are smaller compared with NP spheres. (B) In a similar water–electrolyte–NP–organic-phase system, the NPs have migrated to the aqueous phase, and light propagates through the system because the aqueous phase and organic phase are transparent. *Based on the work of M.E. Flatté, A.A. Kornyshev, M. Urbakh, Electrovariable nanoplasmonics and self-assembling smart mirrors, J. Phys. Chem. C. 114 (2010) 1735, reproduced with permission by the American Chemical Society [54].*

reflection and refraction at the interface are calculated from the given difference between the refractive indices of the oil and aqueous phase.

Fig. 3.12 shows an interfacial mirror consisting of metallic gold NP exhibiting high reflectance and electrical conductance. The particles were self-assembled at a heptane 1,2-dichloroethane–water, liquid–liquid interface, with the highest reflectance exhibited for gold NP films consisting of 60-nm diameter NP and monolayer surface coverage. Interestingly, the scanning electrochemical microscopy revealed that a transition occurs from an insulating to a conducting electrical material upon reaching a surface coverage at least equivalent to the formation of a single monolayer [55]. When red-colored gold nanospheres are used for interfacial migrations and droplet formation, the relative redness between the area of the droplets and the solvent can show the relative concentration of NPs between the droplet and the solvent.

Another work on interfacial mirrors of plasmonics was published by Smirnov and coworkers [56]. They developed a theory based on the Fermi level equilibration of two redox species in immiscible solutions that are separated, but both of them are in contact with an interfacial film of metallic NP. These soft interfaces were half a monolayer of mirror-like nanofilms of gold NP and were characterized by ion transfer voltammetry. The electric polarization of these interfaces can be controlled readily, so the Fermi level (which is the total chemical potential of the electrons in the interfacial gold NP film) can also be manipulated.

"Metamaterials" may also be synthesized through mirrors of metallic NP, which have formed a perfect metallic film. Sikdar and coworkers presented a theoretical study via a five-layer stack model and calculated its reflectance, transmittance, and absorbance spectra through dipolar quasi-static approximations, image forces, and effective medium theory. Their investigations covered the visible spectrum range of 400–900 nm [57a]. The Stellacci research team [57b] presented a synthetic route for the fabrication of ultrathin freestanding NP membranes. The membranes consisted of gold NPs capped with trimethoxysilane ligands, and the mechanism was based on the interfacial assembly in an oil–water mixture. The particles were transferred to the liquid–liquid interface simply upon shaking the solutions. Through a crosslinking of the Si–O–Si bonds, a network was formed, and this led to the formation of ultrathin membranes of only a few NP layers in thickness and a few cm^2 in area. A comprehensive analysis of the optical properties of gold NP assemblies in liquid–liquid interfaces can be found in the work by Paget and coworkers. They carried out studies assuming that core-shell Au@SiO_2 NPs are located predominantly in the aqueous phase. They presented a theoretical model on dipole approximation [58].

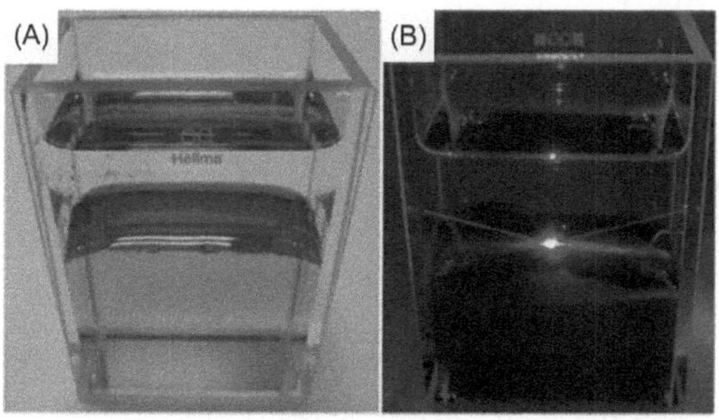

Figure 3.12 (A) A mirror consisting of gold NP (average diameter, 60 nm) synthesized from a heptane–1,2-dichloroethane–water interface. The gold concentration is equivalent to a surface coverage of 1.6 monolayers. (B) Under irradiation with green laser light (532 nm), the highly reflective metallic nature of the mirror is revealed. *Reproduced with permission from P.-P. Fang, S. Chen, H. Deng, M.D. Scanlon, F. Gumy, H.J. Lee, et al., Conductive gold nanoparticle mirrors at liquid/liquid interfaces, ACS Nano 7 (2013) 9241. ©2013, American Chemical Society [55].*

3.7 Redox reactions from functionalized nanomaterials through liquid–liquid interfaces

The synthesis of surface functionalized inorganic NP has been achieved effectively using two-phase systems. Examples are the organophilic, oleic acid-capped, iron oxide maghemite NPs that were synthesized by Bourlinos and coworkers through a toluene–water emulsion. This is a highly effective facile method that leads to an excellent coverage of the NP surface, practically superior than other methods, including the high-temperature synthesis of iron oxides through an in situ solid-state reaction, as described by Zboril and coworkers [59,60].

Other methods include the layer-by-layer approach for multicomponent, multifunctional systems. Kim and coworkers synthesized, through a layer-by-layer method, oleic acid–coated Fe_3O_4, tetraoctylammonium bromide–palladium NP, and oleic acid–TiO_2 NP linked with dendrimer ligands. These compounds demonstrated good performance toward the oxidation of 3,3″,5,5″-tetramethylbenzidine and as photocatalysts in aqueous media and in the Sonogashira coupling reaction with a very high yield (99%) in organic media [61].

Toth and coworkers demonstrated the electrochemical reactivity of carbon nanotube (CNT) films at a polarizable organic–water interface [62] and of functionalized graphene [63,64]. This process utilizes the electrons that are generated through oxidation of 1,10-dimethylferrocene (DMFc) to $DMFc^+$, and they are transferred using the assembled CNT layer. The authors studied the interfacial deposition through a series of electrochemical methods, such as cyclic voltammetry, in situ Raman spectroscopy coupled with electrochemistry, and electrochemical impedance spectroscopy. They demonstrated that the redox couples dope and charge the nanotubes.

Graphene, fullerenes, and CNT (Fig. 3.13) have provided a whole new perspective and direction in materials science due to their unique properties. However, many applications require assembly of carbon nanomaterials into larger and ordered structures. To that end, liquid–liquid interfaces utilizing the high-energy interface between the two liquids can mediate the arrangement and assembly of particles. In these systems, initially particles are dispersed in one of the two bulk liquid phases by reducing the overall energy of the system [65].

In recent works, similar interfacial systems have been applied to the formation of CNT films for electrochemical applications [66] or for composite structures, as demonstrated by Zarbin and coworkers, to form

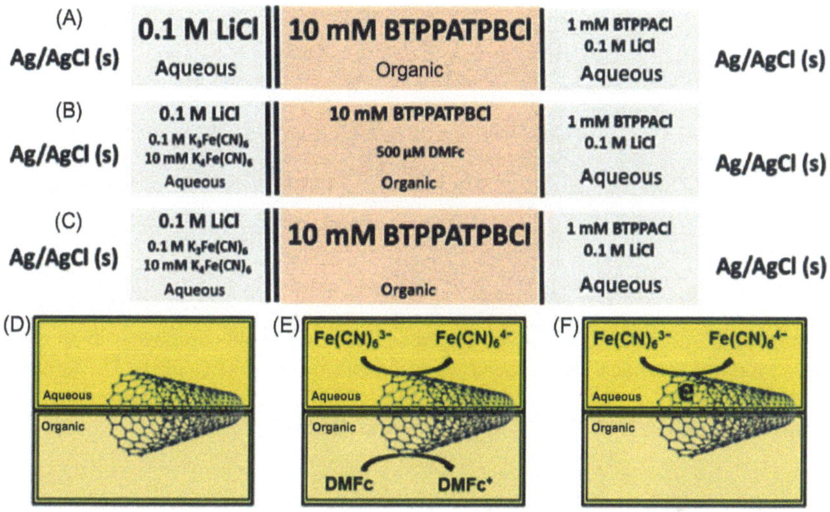

Figure 3.13 Liquid-liquid pseudocapacitors by the utilization of carbon nanotubes. In the figure can be seen the cell compositions for (A) The supporting blank (B) The aqueous and organic redox couple. (C) Only the aqueous redox couple configurations (D-F). Are the respective schematic representations. *Reproduced with permission from P.S. Toth, A.N.J. Rodgers, A.K. Rabiu, D. Ibañez, J.X. Yang, A. Colina et al., Interfacial doping of carbon nanotubes at the polarisable organic/water interface: a liquid/liquid pseudo-capacitor, J. Mater. Chem. A 4 (2016) 7365. ©2016, Royal Society of Chemistry [62].*

composite structures with conductive polyaniline polymers [67]. In all of these systems, the electrolyte is added to one of the two liquid phases, and electrons and ions can contribute to the measured current. This current flow is generated through electron transfer, ion transfer, and facilitated ion transfer between the immiscible electrolyte solutions. This phenomenon has led to the enormous research activity on heterogeneous reductions for potential application in hydrogen evolution and oxygen reduction reactions (ORRs) with organophilic reducing agents [68]. When CNT and graphite oxide have been used as support for the catalysts Mo_2C and MoS_2, they improved catalytic activity due to improved electron transfer from the lipophilic reducing agent to the interfacial catalyst [69a]. Other interfacial charge transfer processes are the aqueous phase versus the interface of room temperature ionic liquids. Potential applications include proton-coupled electron transfer, fuel cells, and hydrogen storage, where ionic liquids are used as aprotic solvents [69b–e].

Figure 3.14 Interfacial assembly of graphene-noble metal NP composites in an organic–aqueous interface. *Reproduced with permission from P.S. Toth, Q. Ramasse, M. Velicky, R.A.W. Dryfe, Functionalization of graphene at the organic/water interface, Chem. Sci. 6 (2015) 1316. ©2015, Royal Society of Chemistry [63].*

In another work, Toth and coworkers reported the deposition of noble metal (palladium, gold) NP on graphene layers that had been synthesized through chemical-vapor deposition in a monolayer (Fig. 3.14). In liquid–liquid interface assemblies, graphene has been reported to form macroscopic films through self-assembly and stacking of individual layers. In this work, layers were transferred using a typical method for the transfer of graphene monolayers with PMMA. Then, attachment of the noble metal NP took place using electrochemical methods. The as-synthesized graphene-based metal nanoclusters were characterized using AFM, electron microscopy, and Raman spectroscopy. For electrochemical deposition of palladium NP on graphene layers, a spontaneous redox process between the metal precursor and organic-phase electron donor is the medium step. It appeared that metal deposition proceeded through the interface and in the presence of graphene layers, and then it occurred preferentially on graphene. The volume ratio of dichloroethane and 5-nonanone was 1:4 to afford rapid dissolution of poly(methyl)methacrylate. The electrochemical step that was described earlier was used only for palladium. For the synthesis of gold particles, the precursor salt, $AuCl_4$, can be transferred from the aqueous solution to the organic phase. The electron donor in the organic phase was DMFc, which is responsible for the reduction of the noble metal cation salts. Supporting electrolytes were 0.1 M LiCl and 10 mM *bis*(diphenyl phosphoranylidene) ammonium tetrakis(4-chlorophenyl) borate in the aqueous and organic phases, respectively [63–65]. The redox process between the palladium chloride and decamethylferrocene (DecMFc) follows the mechanism outlined here.

Another electrochemically driven process at a liquid–liquid boundary has been reported by Rodgers and coworkers. It involves oxygen reduction at the interface of two immiscible electrolyte solutions, which would be of interest for fundamental studies and industrial applications such as sensors, nuclear waste treatment, biomimetics, and in vivo charge and mass transport [70–72]. The authors used graphene assembled through interfaces to demonstrate the importance of simple electron conductivity for oxygen reduction and a bipolar setup to elucidate the homogeneous/heterogeneous nature of the oxygen-reduction and hydrogen-evolution reactions mediated through interfacially assembled graphene [73].

In a similar interfacial process, Peljo and coworkers [74] presented an ORRs catalyzed by cofacial metalloporphyrins at a 1,2-dichlorobenzene–water interface (Fig. 3.15). They used two organophilic electron donors: DMFc and tetrathiafulvalene. The products of the reaction were water and hydrogen peroxide. It was found that the mediator dictates the selectivity significantly, with DMFc and porphyrins having a catalytic role over the decomposition and further reduction of hydrogen peroxide. In addition to the experimental work, the authors undertook density functional theory calculations, which suggested that the biscobaltporphyrin 4,5-*bis*[5-(2,8,13,17-tetraethyl-3,7,12,18-tetramethylporphyrinyl)]-9,9-dimethylxanthene [Co_2(DPX)] would catalyze the ORRs toward hydrogen peroxide only if the oxygen is bound on the "exo" side of the catalyst. In contrast, a four-electron reduction process will take place if the oxygen is bound on the "endo" side of the molecule.

3.8 Photonic crystals

The tail of a peacock is an example of the unparalleled ability of nature to create exquisite ordered structures with unique functionalities and beauty. The intense and fiery blue and green on the tail originates from a nanoscale structure of the feathers that allows the light of a specific wavelength to be amplified and blocks the light other wavelengths. The color that arises from photonic assemblies is called "structural color" and inspired the roman poet, Lucretius (99–55 BCE) [75] There is a wealth of photonic nanostructures that have been synthesized, including the inverse opal arrays of Joanna Aizenberg [76,77] and amphiphilic magnetic NP [78]. For example, in the work by Dorvee and coworkers, controlled manipulation of amphiphilic, superparamagnetic, and iron oxide fluid droplets were presented [78]. This manipulation involves, in general, the control

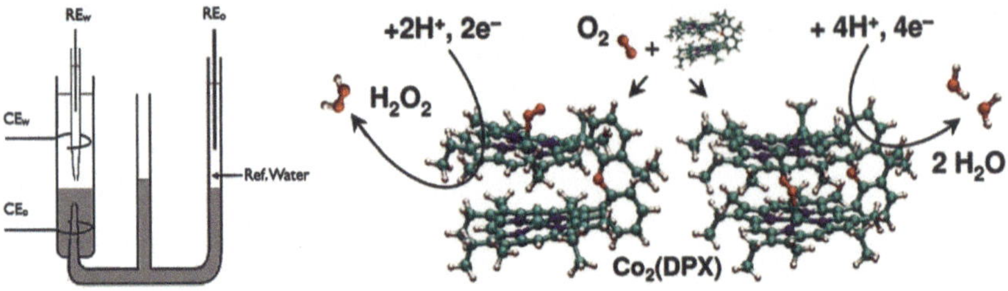

Figure 3.15 The electrochemical cell employed by Peljo and coworkers. *Reproduced with permission from P. Peljo, L. Murtomäki, T. Kallio, H.-J. Xu, M. Meyer, C.P. Gros, et al., Biomimetic oxygen reduction by cofacial porphyrins at a liquid–liquid interface, J. Am. Chem. Soc. 134 (2012) 5974. ©2012, American Chemical Society [74].*

of small volumes of fluids. It is an important challenge in microfluidics [79] since it represents a key requirement for various applications and assays [80].

In previous works, one-dimensional photonic crystals have been constructed by using porous silicone templates. These nanocrystals have demonstrated an amphiphilic character. Due to their amphiphilic behavior, if placed in a two-phase liquid–liquid system, for example, dichloromethane–water, they tend to accumulate and align spontaneously at the interface. In the field of magnetic materials, Dorvee and coworkers demonstrated that superparamagnetic NPs of iron oxide (the particles were in the oxidation state and crystal structure of magnetite) could be incorporated into a porous nanostructure. Also, they could be manipulated through an external magnetic field, with the optical reflectivity spectrum of the fluid displaying a peak, which is used to identify the droplet, with the latter also used as a medium for carrying out a chemical reaction. For the synthesis of the droplet, the particle film is hydrosilylated in the presence of 1-dodecene [81]. In general, the liquid–liquid interfaces have been employed widely for the synthesis of mesoporous materials (see Li and coworkers for a review of published work [82]).

3.9 Biomimetics, molecular recognition, and nanomachines at a liquid–liquid interface

Biomimetics is the attempt to copy the self-assembly processes of nature and biological systems and apply them to laboratory-constructed systems. It is with respect to the biological systems that we initially discussed the notion of self-assembly of molecules and particles and the construction of functional nanomachines [83]. Chapter 1, Hydrophilic and Organophilic Systems in Nature, discussed the structure and the function of the biological membrane and the role of amphiphilic molecules on its structure, with a focus on ion transfer. Besides the permeability of ions and small molecules, several other types of reactions and processes take place at membranes. For example, ORR in the cell respiration chain. It is catalyzed through enzymes with porphyrin subunits as active centers, which are attached to the hydrophobic part of the lipid comprising the bilayer of mitochondrial membranes. There have been many reports on the catalytic activity of cobalt phthalocyanine for oxygen reduction [84,85] as demonstrated by measurements of the catalyst attached on a carbon electrode [86]. The effect has been demonstrated in organic media as well; organic acids were used and metalloporphyrins catalyzed oxygen reduction to hydrogen peroxide by ferrocene in acetonitrile [87] or benzonitrile [88,89]. These porphyrins have a similar structure and resemble the cytochrome *c* oxidase, which dictates the selective four-electron reduction of molecular oxygen in nature.

Stockmann and coworkers studied the interaction between protons and lithium cations with phospholipids in an electrochemical system. The reactions took place at a water–chloroform interface [90]. Through the fabrication of this interface, they attempted to mimic biological membranes in which ions are transported throughout phospholipid membranes. This biomimetic scheme can be expanded to include larger alkali metal ions. Hence, interfacial electrochemistry can be regarded to be a man-made, laboratory-based analogue of phospholipid biological membranes with ion flow and different current density. Fig. 3.16 is a schematic representation of the experiments of Stockmann and coworkers. DecMFc was used as the "sacrificial" electron donor, and their cell was used for the reduction of oxygen and evolution of hydrogen (i.e., they were used as a solar fuel generator).

Two-phase systems such as water–1,2-dichloroethane have been used as laboratory-based examples for a transfer reaction of biomolecules. They can be used to elucidate the interaction of ions with porous structures at the nanoscale, a transfer practically similar to the one taking place at biological cells [90b–e]. For example, insulin has been detected at liquid–liquid microinterfaces with bovine albumin; interfacial probes have been used for the electrochemical detection of acetylcholine, tryptamine, and serotonin through ionic transfer; dielectric layers through protein adsorption have been formed at biomimetic interfaces.

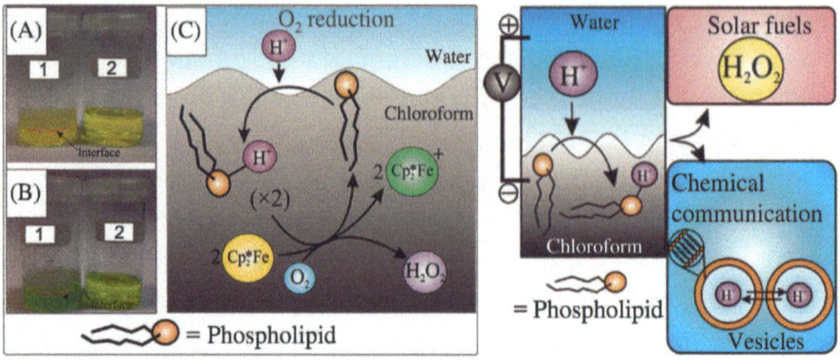

Figure 3.16 An aqueous–1,2-dichloromethane interfacial system with a catalytic amount of the phospholipid (the concentration of the phospholipid 1,2-dimyristoyl-sn-glycero-3-phosphocholine was approximately 4 μM). (A) Photographs of shake-flask experiments performed at time 0. (B) A photograph after 24 hours. B1 and B2 contained an aqueous phase with (B1) and without acid and tetramethylammonium tetraphenylborate (B2). Even with such a small amount of the phospholipid, oxygen reduction takes place. *Reproduced with permission by T.J. Stockmann, J.-M. Noel, A.A. Hassan, C. Combellas, F. Kanou. Facilitated lewis acid transfer by phospholipids at a (water|CHCl3) liquid|liquid interface toward biomimetic and energy applications, J. Phys. Chem. C 120 (2016) 11977. ©2016, American Chemical Society [90a].*

3.10 Gyroid polymeric nanostructures from interfaces

Another interesting class of polymeric nanomaterials that may assemble in a wealth of microdomains and structures are the gyroid block copolymers [91]. The behavior of various diblock or triblock copolymers in oil–water interfaces has been studied. It was found that triblock (ABA type) copolymers cause interface deformations depending on the ratio of the different block subcomponents [92]. They may form a wealth of nanostructures in various phases (including helical) and conventional nanostructures (e.g., spheres, hexagonally packed cylinders, double gyroid structures, lamellae arrays). Gyroid materials have found applications as novel templates, photonics, and in fabrication of metamaterials [93]. A typical copolymer that may follow from these self-assembled pathways is the polystyrene-*block*-poly(L-lactide).

One of the most complex geometrical structures generated by block copolymers is gyroid. The gyroid structure is an infinitely interconnected and complex, with curls in various styles that leave a network of holes that were discovered first by the mathematician Alan Schoen in 1970 [94]. It is an exquisite structure responsible for the intense iridescent color from the multilayer-based biological entities such as the fruits of *Pollia condensata* and the famous iridescence of the *Blue morpho* butterfly. This iridescence is caused by the Bragg reflection of helicoidally stacked multilayers of cellulose microfibrils [95,96].

The gyroid structure generated by certain triblock copolymers appears to consist of three interconnected units, but their mutual repulsion leads to a complex relationship and nanoscale arrangements. Jeremy Baumberg, by following self-assembly procedures, synthesized the first nanoscale optical metamaterial in 3D using gyroid structures [96].

Asghar and coworkers developed a methodology that predicts and controls the lattice parameters of hexagonal and gyroid mesoporous materials. These materials are formed through liquid crystal templating and are predicted as a function of their water/surfactant/oil volume fractions. They applied their simulations to the lamellar (L_α), hexagonal (H_1), and gyroid bicontinuous cubic (V_1) mesophases. The systems that were studied were the binary polyoxyethylene cetyl ether ($C_{16}EO_{10}$)−water system and the ternary polyoxyethylene cetyl ether−hexadecane−water system. Also, they synthesized mesoporous platinum using hexagonal and gyroid phases as templates [97]. Gyroid structures [97−99] have been utilized recently in an ingenious way by the research teams of Uli Wiesner, Francis DiSalvo, and Bruce Van Dover as templates for the synthesis of niobium nitride superconductors from a chiral ABC triblock terpolymer [98−100]. The Cornell University consortium used two triblock terpolymers, the poly(isoprene)-*block*-poly(styrene)-*block*-poly(ethylene oxide) and the nd poly(isoprene)-*block*-poly(styrene)-*block*-poly(ethylene oxide) for the direct synthesis of the sol gel−derived niobium oxide (Nb2O5) as a three-dimensional network structure.

3.11 Conclusion

Liquid−liquid interfaces and micellar assemblies have been shown to be powerful systems for the self-organization of materials. Also, they have been used for the applications such as efficient media for electrochemical reactions, biomimetic processes, and construction of reflective interfacial mirrors through densely packed plasmonic nanomaterials. These noble metal arrays possess unique optical properties and with respect to their surface functionalities and can form various optically and catalytically functional structures. Through fabrication of novel Janus particles [101], gyroid polymer structures and noble metal plasmonics with different sizes, shapes, and functionalities can be developed further. Of particular interest is the combination of 2D materials, such as graphene or perhaps other semiconductors, with noble metal NP for catalytic and electrochemical applications. Graphene, CNTs, and palladium composites have been implemented through interfacial electrochemistry. The advent of the plasmonic mirrors and the metamaterials assemblies will pave the way for the fabrication of unique systems for bespoke applications. The development of novel, flexible, and light responsive organic ligands can offer new and exciting possibilities for responsive, functional 2-D arrays. In Chapter 4, Synthesis and Generation of Polymeric Materials From Interfaces, the emphasis now will be given on polymeric materials, especially those possessing electronic conductivity properties [102].

References

[1] B. Petkova, S. Tcholakova, M. Chenkova, K. Golemanov, N. Denkov, D. Thorley, et al., Foamability of aqueous solutions: role of surfactant type and concentration, Adv. Coll. Int. Sci. 276 (2020) 102084.

[2] (a) S. Shi, T.P. Russell, Nanoparticle assembly at liquid–liquid interfaces: from the nanoscale to mesoscale, Adv. Mater. 30 (2018) 1800714.
(b) B. Smit, P.A.J. Hilbers, K. Esselink, L.A.M. Rupert, N.M. van Os, Structure of a water/oil interface in the presence of micelles: a computer simulation study, J. Phys. Chem. 95 (1991) 6361.
(c) B.J. Park, D. Lee, Spontaneous particle transport through a triple-fluid phase boundary, Langmuir 29 (2013) 9662.

[3] R.G. Larson, Nucleation of nitrogen: experiment and theory, J. Phys. Chem. 91 (1989) 2479.

[4] E.M. Lee, R.K. Thomas, J. Penfold, R.C. Ward, Structure of aqueous decyltrimethylammonium bromide solutions at the air water interface studied by the specular reflection of neutrons, J. Phys. Chem. 93 (1989) 381.

[5] D.W.R. Gruen, S. Marcelja, Spatially varying polarization in water—a model for the electric double-layer and the hydration force, J. Chem. Soc. Faraday Trans. 79 (1983) 225.

[6] S. Marcelja, N. Radic, Repulsion of interfaces due to boundary water, Chem. Phys. Lett. 42 (1976) 129.

[7] D.M. LeNeveu, R.P. Rand, V.A. Parsegian, Measurement of forces between lecithin bilayers, Nature 259 (1976) 601.

[8] S. Wada, K. Fujiwara, H. Monjushiro, H. Watarai, Measurement of circular dichroism spectra of liquid/liquid interface by centrifugal liquid membrane method, Anal. Sci. 20 (2004) 1489.

[9] (a) E.A. Simister, E.M. Lee, J.R. Lu, R.K. Thomas, R.H. Ottewill, A.R. Rennie, et al., Adsorption of ammonium perfluorooctanoate and ammonium decanoate at the air/solution interface, J. Chem. Soc. Faraday Trans. 88 (1992) 3033.
(b) E.H. Lucassen-Reynders, J. Lucassen, D. Giles, Surface and bulk properties of mixed anionic/cationic surfactant systems i. equilibrium surface tensions, J. Coll. Int. Sci 81 (1981) 150.

[10] G.-H. Li, F. Li, L.-Q. Zheng, H.-L. Wang, Fluorescence-probe study of anionic/cationic surfactant systems 2. Alkylsulfonate/alkyltrimethylammonium bromide, Colloids Surf. A 76 (1993) 257.
[11] Y.-L. You, L.-S. Hao, Y.-Q. Nan, Phase behavior and viscous properties of cetyltrimethylammonium bromide and sodium dodecyl sulfonate aqueous mixtures, Colloids Surf. A 335 (2009) 154.
[12] B.Y. Zhu, M.J. Rosen, Synergism in binary mixtures of surfactants: IV. Effectiveness of surface tension reduction, J. Colloid Interface Sci. 99 (1984) 435.
[13] M.T. Yatcilla, K.L. Herrington, L.L. Brasher, E.W. Kaler, S. Chiruvolu, A. Zasadzinski, Phase behavior of aqueous mixtures of cetyltrimethylammonium bromide (CTAB) and sodium octyl sulfate (SOS), J. Phys. Chem. 100 (1996) 5874.
[14] C. Tondre, C. Caillet, Properties of the amphiphilic films in mixed cationic/anionic vesicles: a comprehensive view from a literature analysis, Adv. Colloid Interface Sci. 93 (2001) 115.
[15] P.K. Yuet, D. Blankschtein, Effect of surfactant tail-length asymmetry on the formation of mixed surfactant vesicles, Langmuir 12 (1996) 3819.
[16] M. Antonietti, T. Forster, Vesicles and liposomes: a self-assembly principle beyond lipids, Adv. Mater. 15 (2003) 1323.
[17] (a) V. Guida, Thermodynamics and kinetics of vesicles formation processes, Adv. Colloid Interface Sci. 161 (2010) 77.
(b) G. Battaglia, A.J. Ryan, The evolution of vesicles from bulk lamellar gels, Nat. Mat. 4 (2005) 869.
(c) A.J. Ryan, J.P.A. Fairclough, I.A. Hamley, S.-M. Mai, C. Booth, Chain folding in crystallizable block copolymers, Macromolecules 30 (1997) 1723.
[18] (a) T. Homma, K. Harano, H. Isobe, E. Nakamura, Preparation and properties of vesicles made of nonpolar/polar/nonpolar fullerene amphiphiles, J. Am. Chem. Soc. 133 (2011) 6364.
(b) P. Bairi, K. Minami, J.P. Hill, W. Nakanishi, L.K. Shrestha, C. Liu, et al., Supramolecular differentiation for construction of anisotropic fullerene nanostructures by time-programmed control of interfacial growth, ACS Nano. 10 (2016) 8796.
[19] O.J. Cayre, V.N. Paunov, Langmuir. 20 (22) (2004) 9594–9599.
[20] (a) R.M. Gorgoll, T. Tsubota, K. Harano, E. Nakamura, Cooperative self-assembly of gold nanoparticles on the hydrophobic surface of vesicles in water, J. Am. Chem. Soc. 137 (2015) 7568.
(b) V. Müller, F. Shao, M. Baljozovic, M. Moradi, Y. Zhang, T. Jung, et al., Structural characterization of a covalent monolayer sheet obtained by two-dimensional polymerization at an air/water interface, Angewandte Chemie Int. Edition 56 (2017) 15262–15266.
(c) H. Al-Shehri, T.S. Horozov, V.N. Paunov, Adsorption of carboxylic modified latex particles at liquid interfaces studied by the gel trapping technique, Soft Matter. 10 (34) (2014) 6433–6441.
[21] P.W.K. Rothemund, Folding DNA to create nanoscale shapes and patterns, Nature 440 (2006) 297.
[22] S. Srivastava, D. Nykypanchuk, M. Fukuto, J.D. Halverson, A.V. Tkachenko, K.G. Yager, et al., Two-dimensional DNA-programmable assembly of nanoparticles at liquid interfaces, J. Am. Chem. Soc. 136 (2002) 8323.
[23] S. Srivastava, D. Nykypanchuk, M. Fukuto, O. Gang, Tunable nanoparticle arrays at charged interfaces, ACS Nano 8 (2002) 9857.
[24] K. Jiang, S. Ma, Y. Zhang, X. Han, Formation of square prism-shaped poly(o-phenylenediamine) fibers triggered by high ionic strength, RSC Adv. 6 (2016) 21895.

[25] D.D. Sawall, R.M. Villahermosa, R.A. Lipeles, A.R. Hopkins, Interfacial polymerization of polyaniline nanofibers grafted to Au surfaces, Chem. Mater. 16 (2004) 1606.

[26] H. Zhang, W. Wang, N. Hagen, I. Kuzmenko, M. Akinc, A. Travesset, et al., Self-assembly of DNA functionalized gold nanoparticles at the liquid-vapor interface, Adv. Mater. Interfaces 3 (2016) 1600180.

[27] W.H. Binder, Supramolecular assembly of nanoparticles at liquid–liquid interfaces, Angew. Chem. Int. Ed. 44 (2005) 5172.

[28] S.G. Booth, R.A.W. Dryfe, Assembly of nanoscale objects at the liquid/liquid interface, J. Phys. Chem. C. 119 (2015) 23295.

[29] G.M. Whitesides, M. Boncheva, Beyond molecules: self-assembly of mesoscopic and macroscopic components, Proc. Natl Acad. Sci. U.S.A. 99 (2002) 4769.

[30] P. Pieranski, Two-dimensional interfacial colloidal crystals, Phys. Rev. Lett. 45 (1980) 569.

[31] B. Pansu, P. Pieranksi, L. Strzelecki, Thin colloidal crystals: a series of structural transitions, J. de. Phys. Paris. 44 (1983) 531.

[32] A. Saha, V.T. John, A. Bose, In situ assembly of hydrophilic and hydrophobic nanoparticles at oil–water interfaces as a versatile strategy to form stable emulsions, ACS Appl. Mater. Interfaces 7 (2015) 21010.

[33] G.W. Githaiga, A.W. Woodward, A.R. Morales, M.V. Bondar, K.D. Belfield, Photophysical and computational analysis of a symmetrical fluorene-based janus dione derivative, J. Phys. Chem. C. 119 (2015) 21053.

[34] T.M. Ruhland, A.H. Gröschel, N. Ballard, T.S. Skelhon, A. Walther, A.H.E. Müller, et al., Influence of Janus particle shape on their interfacial behavior at liquid–liquid interfaces, Langmuir 29 (2013) 1388.

[35] (a) L. Sheng, H. Chen, W. Fu, Z. Li, Janus silica hollow spheres prepared via interfacial biosilicification, Langmuir 31 (2015) 11964.
(b) A. Walther, X. Andre, M. Drechsler, V. Abetz, A.H.E. Müller, Janus discs, J. Am. Chem. Soc. 129 (2007) 6187.
(c) T. Ruhland, A. Gröschel, A. Walther, A.H.E. Müller, Janus cylinders at liquid–liquid interfaces, Langmuir 27 (2011) 9807.

[36] B.J. Park, E.M. Furst, Fabrication of unusual asymmetric colloids at an oil – water interface, Langmuir 26 (2010) 10406.

[37] H. Duan, D. Wang, D.G. Kurth, H.M.D. Mohwald, Directing self-assembly of nanoparticles at water/oil interfaces, Angew. Chem. Int. Ed. 116 (2004) 5757.

[38] H. Gu, Z. Yang, J. Gao, C.K. Changm, B. Xu, Heterodimers of nanoparticles: formation at a liquid – liquid interface and particle-specific surface modification by functional molecules, J. Am. Chem. Soc. 127 (2005) 34.

[39] (a) P.V. Petkov, K.D. Danov, P.A. Kralchevsky, Surface pressure isotherm for a monolayer of charged colloidal particles at a water/nonpolar-fluid interface: experiment and theoretical model, Langmuir 30 (2014) 2768.
(b) P.A. Kralchevsky, D.D. Krassimir, V.P. Petkov, Soft electrostatic repulsion in particle monolayers at liquid interfaces: surface pressure and effect of aggregation, Philosophic. Trans. Royal Soc. A 374 (2016) 20150130.
(c) R. Vecchione, G. Iaccarino, P. Bianchini, R. Marotta, F. D'autilia, V. Quaqliariello, et al., Ultrastable liquid–liquid interface as viable route for controlled deposition of biodegradable polymer nanocapsules, Small 12 (2016) 3005.

[40] N. Glaser, D.J. Adams, A. Boker, G. Krausch, Janus particles at liquid – liquid interfaces, Langmuir 22 (2006) 5227.

[41] (a) K.L. Kelly, E. Coronado, L.L. Zhao, G.C. Schatz, The optical properties of metal nanoparticles: the influence of size, shape, and dielectric environment, J. Phys. Chem. B 107 (2002) 668.

(b) M.B. Ross, C.A. Mirkin, G.C. Schatz, Optical properties of one-, two-, and three-dimensional arrays of plasmonic nanostructures, J. Phys. Chem. C 120 (2016) 816.
(c) M. Lee, M. Xia, P. Ming, J. Bum, Materials 9 (2016) 138.

[42] P.A. Sanchez, E.S. Pyanzina, E.V. Novak, J.J. Cerda, T. Sintes, S.S. Kantorovich, Supramolecular magnetic brushes: the impact of dipolar interactions on the equilibrium structure, Macromolecules 48 (2015) 7658.

[43] H. Lehle, M. Oettel, Importance of boundary conditions for fluctuation-induced forces between colloids at interfaces, Phys. Rev. E 751 (2007) 011602.

[44] V.A. Turek, M.P. Cecchini, J. Paget, A.R. Kucernak, A.A. Kornyshev, J.B. Edel, Plasmonic ruler at the liquid–liquid interface, ACS Nano 6 (2012) 7789.

[45] (a) S. Barman, G.F. Christopher, Simultaneous interfacial rheology and microstructure measurement of densely aggregated particle laden interfaces using a modified double wall ring interfacial rheometer, Langmuir 30 (2014) 9752.
(b) L. Wu, Z. Liao, M. Liu, X. Yin, X. Li, M. Wang, et al., Fabrication of non-spherical pickering emulsion droplets by cyclodextrins mediated molecular self-assembly, Colloids Surf. A 490 (2016) 163.
(c) P. Bertsch, M. Arcari, T Geue, T. Geue, R. Mezzenga, G. Nyström, et al., Designing cellulose nanofibrils for stabilization of fluid interfaces, Biomacromolecules 20 (12) (2019) 4574.

[46] S. Pothorszky, D. Zámbó, T. Deák, A. Deák, Detecting patchy nanoparticle assembly at the single-particle level, Nanoscale 8 (2016) 3523.

[47] J. Che, K. Park, C.A. Grabowski, A. Jawaid, J. Kelley, H. Koerner, et al., Preparation of ordered monolayers of polymer grafted nanoparticles: impact of architecture, concentration, and substrate surface energy, Macromolecules 49 (2016) 1834.

[48] (a) N. Xue, S. Wu, S. Sun, D. Quéré, Q. Zheng, Strongly metastable assemblies of particles at liquid interfaces, Langmuir 30 (2014) 14712.
(b) M. Sabapathy, V. Kollabattula, M.G. Basavaraj, E. Mani, Visualization of the equilibrium position of colloidal particles at fluid-water interfaces by deposition of nanoparticles, Nanoscale 7 (2015) 13868.
(c) L. Costa, G. Li-Destri, N.H. Thomson, O. Konovalov, D. Pontoni, Real space imaging of nanoparticle assembly at liquid–liquid interfaces with nanoscale resolution, Nano Lett. 16 (2016) 5463.
(d) D.J. Muller, AFM: a nanotool in membrane biology, Biochemistry 47 (2008) 7986.
(e) A.M. Jubb, W. Hua, H.C. Allen, Acc. Chem. Res. 45 (2012) 110–119.

[49] (a) J. He, P. Kanjanaboos, N.L. Frazer, A. Weis, X.-M. Lin, H.M. Jaeger, Fabrication and mechanical properties of large-scale freestanding nanoparticle membranes, Small 6 (2010) 1449.
(b) K.E. Mueggenburg, X.-M. Lin, R.H. Goldsmith, H.M. Jaeger, Elastic membranes of close-packed nanoparticle arrays, Nat. Mat. 6 (2007) 656.
(c) K.M. Salerno, D.S. Bolintineanu, J.M.D. Lane, G.S. Crest, High strength, molecularly thin nanoparticle membranes, Phys. Rev. Lett. 113 (2014) 258301.
(d) Y.-G. Jung, C.-R. Lee, H.-J. Lee, J.W. Chang, H.-Y. Lee, Soft Matter 15 (2019) 3949–3956.

[50] X. Fan, W. Zheng, D.J. Singh, Light Sci. Appl. 3 (2014) e179.

[51] D. Yogev, S. Efrima, Novel silver metal liquidlike films, J. Phys. Chem. 92 (1988) 5754.

[52] A.A. Kornyshev, M. Marinescu, J. Paget, M. Urbakh, Reflection of light by metal nanoparticles at electrodes, Phys. Chem. Chem. Phys. 14 (2012) 1850.

[53] M.E. Flatté, A.A. Kornyshev, M. Urbakh, Nanoparticles at electrified liquid–liquid interfaces: new options for electro-optics, Faraday Discuss. 143 (2009) 109.

[54] M.E. Flatté, A.A. Kornyshev, M. Urbakh, Electrovariable nanoplasmonics and self-assembling smart mirrors, J. Phys. Chem. C. 114 (2010) 1735.
[55] P.-P. Fang, S. Chen, H. Deng, M.D. Scanlon, F. Gumy, H.J. Lee, et al., Conductive gold nanoparticle mirrors at liquid/liquid interfaces, ACS Nano 7 (2013) 9241.
[56] E. Smirnov, P. Peljo, M.D. Scanlon, H.H. Girault, Interfacial redox catalysis on gold nanofilms at soft interfaces, ACS Nano 9 (2015) 6565.
[57] (a) D. Sikdar, S.B. Hasan, M. Urbakh, J.B. Edel, A.A. Kornyshev, Unravelling the optical responses of nanoplasmonic mirror-on-mirror metamaterials, Phys. Chem. Chem. Phys. 18 (2016) 20486.
(b) B. Le Ouay, S. Guldin, Z. Luo, S. Allegri, F. Stellacci, Freestanding ultrathin nanoparticle membranes assembled at transient liquid−liquid interfaces, Adv. Mat. Interfac. 3 (2016) 1600191.
[58] J. Paget, V. Walpole, M.B. Jorquera, J.B. Edel, M. Urbakh, A.A. Kornyshev, et al., Optical properties of ordered self-assembled nanoparticle arrays at interfaces, J. Phys. Chem. C. 118 (2014) 23264.
[59] P. Dallas, A.B. Bourlinos, D. Niarchos, D. Petridis, Synthesis of tunable sized capped magnetic iron oxide nanoparticles highly soluble in organic solvents, J. Mater. Sci. 42 (2007) 4996.
[60] R. Zboril, A. Bakandritsos, M. Mashlan, V. Tzitzios, P. Dallas, C. Trapalis, et al., One-step solid state synthesis of capped γ-Fe2O3 nanocrystallites, Nanotechnology 19 (2008) 095602.
[61] D. Kim, S. Cheong, Y. Gyong, A.S.W. Ryu, J.-K. Kim, J. Cho, Multicatalytic colloids with highly scalable, adjustable, and stable functionalities in organic and aqueous media, Nanoscale 8 (2016) 7000.
[62] P.S. Toth, A.N.J. Rodgers, A.K. Rabiu, D. Ibañez, J.X. Yang, A. Colina, et al., Interfacial doping of carbon nanotubes at the polarisable organic/water interface: a liquid/liquid pseudo-capacitor, J. Mater. Chem. A 4 (2016) 7365.
[63] P.S. Toth, Q. Ramasse, M. Velicky, R.A.W. Dryfe, Functionalization of graphene at the organic/ water interface, Chem. Sci. 6 (2015) 1316.
[64] P.S. Toth, A.N.J. Rodgers, A.K. Rabiu, R.A.W. Dryfe, Electrochemical activity and metal deposition using few-layer graphene and carbon nanotubes assembled at the liquid-liquid interface, Electrochem. Commun. 50 (2015) 6.
[65] (a) D.Y. Wang, H.W. Duan, H. Mohwald, The water/oil interface: the emerging horizon for self-assembly of nanoparticles, Soft Matter 1 (2005) 412.
(b) L.F. Hu, M. Chen, X.S. Fang, L.M. Wu, Oil−water interfacial self-assembly: a novel strategy for nanofilm and nanodevice fabrication, Chem. Soc. Rev. 41 (2012) 1350.
(c) S.U. Pickering, CXCV—the chemistry of bordeaux mixture, J. Chem. Soc., Trans. 91 (1907) 2001.
(d) B.P. Binks, Colloidal particles at liquid interfaces, Phys. Chem. Chem. Phys. 9 (2007) 6298.
(e) B.P. Binks, Particles as surfactants—similarities and differences, Curr. Opin. Colloid Interfac. Sci. 7 (2002) 21.
[66] (a) P. Chaturbedy, H.S.S.R. Matte, R. Voggu, A. Govindaraj, C.N.R. Rao, Self-assembly of C60, SWNTs and few-layer graphene and their binary composites at the organic−aqueous interface, J. Colloid Interfac. Sci. 360 (2011) 249.
(b) P. Asuri, S.S. Karajanagi, J.S. Dordick, R.S. Kane, Directed assembly of carbon nanotubes at liquid − liquid interfaces: nanoscale conveyors for interfacial biocatalysis, J. Am. Chem. Soc. 128 (2006) 1046.
(c) J. Matsui, M. Iko, N. Inokuma, H. Orikasa, M. Mitsuishi, T. Kyotani, et al., Fabrication of densely packed multi-walled carbon nanotube ultrathin films using a liquid−liquid interface, Chem. Lett. 35 (2006) 42.

(d) Y.J. Zhang, Y.F. Shen, D. Kuehner, S.X. Wu, Z.M. Su, S. Ye, et al., Directing single-walled carbon nanotubes to self-assemble at water/oil interfaces and facilitate electron transfer, Chem. Communicat. 4273 (2008).
(e) J. Matsui, K. Yamamoto, T. Miyashita, Assembly of untreated single-walled carbon nanotubes at a liquid–liquid interface, Carbon 47 (2009) 1444.
(f) R. Kudo, J. Matsui, T. Shibata, T. Miyashita, Mol. Cryst. Liq. Cryst. 519 (2010) 9.
(g) C.E. Cava, R.V. Salvatierra, D.C.B. Alves, A.S. Ferlauto, A.J.G. Zarbin, L.S. Roman, Self-assembled films of multi-wall carbon nanotubes used in gas sensors to increase the sensitivity limit for oxygen detection, Carbon 50 (2012) 1953.
[67] (a) R.V. Salvatierra, M.M. Oliveira, A.J.G. Zarbin, One-pot synthesis and processing of transparent, conducting, and freestanding carbon nanotubes/polyaniline composite films, Chem. Mat. 22 (2010) 5222.
(b) V.H.R. de Souza, M.M. Oliveira, A.J.G. Zarbin, Thin and flexible all-solid supercapacitor prepared from novel single wall carbon nanotubes/polyaniline thin films obtained in liquid–liquid interfaces, J. Power Sourc. 260 (2014) 34.
[68] (a) M.A. Mendez, R. Partovi-Nia, I. Hatay, B. Su, P.Y. Ge, A. Olaya, et al., Phys. Chem. Chem. Phys. 12 (2010) 15163.
(b) B. Su, H.H. Girault, Z. Samec, Catalysis in Electrochemistry, John Wiley & Sons, Inc, Hoboken, NJ, 2011, p. 427.
(c) A.N.J. Rodgers, S.G. Booth, R.A.W. Dryfe, Electrochem. Communicat. 47 (2014) 17.
[69] (a) X.J. Bian, M.D. Scanlon, S.N. Wang, L. Liao, Y. Tang, B.H. Liu, et al., Chem. Sci. 4 (2013) 3432.
(b) B.M. Quinn, Z.F. Ding, R. Moulton, A.J. Bard, Langmuir 18 (2002) 1734.
(c) E.A. de Eulate, D.S. Silvester, D.W.M. Arrigan, Angewandte Chemie Int. Ed 54 (2015) 14903.
(d) P.S. Toth, A.K. Rabiu, R.A.W. Dryfe, Electrochem. Communicat. 60 (2015) 153.
(e) D. Ibanez, D. Plana, A. Heras, D.J. Fermın, A. Colina, Electrochem. Communicat. 54 (2015) 14.
[70] F. Reymond, D. Fermín, H.J. Lee, H.H. Girault, Electrochim. Acta 45 (2000) 2647.
[71] P. Vanýsek, L.B. Ramírez, J. Chil. Chem. Soc. 53 (2008) 1455.
[72] A. Trojánek, J. Langmaier, Z. Samec, Electrochem. Commun. 8 (2006) 1475.
[73] A.N.J. Rodgers, R.A.W. Dryfe, ChemElectroChem 3 (2016) 472.
[74] P. Peljo, L. Murtomäki, T. Kallio, H.-J. Xu, M. Meyer, C.P. Gros, et al., Biomimetic oxygen reduction by cofacial porphyrins at a liquid–liquid interface, J. Am. Chem. Soc. 134 (2012) 5974.
[75] P. Forbes, T. Crimsey, NanoScience: Giants of the Infinitesimal, Papadakis, Winterbourne, 2014.
[76] T.A. Singleton, I.B. Burgess, B.A. Nerger, A.G. Hanssens, N. Koay, C.J. Barrett, et al., Photo-tuning of highly selective wetting in inverse opals, Soft Matter 10 (2014) 1325.
[77] B. Hatton, L. Mishchenko, S. Davis, K.H. Sandhage, J. Aizenberg, Proc. Natl Acad. Sci. U.S.A. 107 (2010) 10354.
[78] J.R. Dorvee, A.M. Derfus, S.N. Bhatia, M.J. Sailor, Manipulation of liquid droplets using amphiphilic, magnetic one-dimensional photonic crystal chaperones, Nat. Mater. 3 (2004) 896.
[79] (a) P. Aussillous, D. Quere, Liquid marbles, Nature 411 (2001) 924.
(b) O.D. Velev, B.G. Prevo, K.H. Bhatt, On-chip manipulation of free droplets, Nature 426 (2003) 515.
[80] J. Liu, C. Hansen, S.R. Quake, Solving the "world-to-chip" interface problem with a microfluidic matrix, Anal. Chem. 75 (2003) 4718.

[81] (a) M.P. Stewart, J.M. Buriak, Photopatterned hydrosilylation on porous silicon, Angewandte Chemie Int. Ed. 37 (1998) 3257.
(b) R. Boukherroub, J.T.C. Wojtyk, D.D.M. Wayner, D.J. Lockwood, J. Electrochem. Soc. 149 (2002) 59.
[82] W. Li, Q. Yue, Y. Deng, D. Zhao, Ordered mesoporous materials based on interfacial assembly and engineering, Adv. Mater. 25 (2013) 5129.
[83] E. Roduner, Nanoscopic Materials, Royal Society of Chemistry, London, 2014.
[84] R. Jasinski, Nature 201 (1964) 1212.
[85] R. Boulatov, in: J.H. Zagal, F. Bedioui, J.-P. Dodelet (Eds.), N4-Macrocyclic Metal Complexes, Springer, New York, 2006, p. 1.
[86] J.H. Zagal, M.A. Paez, J.F. Silva, in: J.H. Zagal, F. Bedioui, J.-P. Dodelet (Eds.), N4-Macrocyclic Metal Complexes, Springer, New York, 2006, p. 41.
[87] S. Fukuzumi, S. Mochizuki, T. Tanaka, Efficient reduction of dioxygen with ferrocene derivatives, catalyzed by metalloporphyrins in the presence of perchloric acid, Inorg. Chem. 28 (1989) 2459.
[88] S. Fukuzumi, K. Okamoto, C.P. Gros, R. Guilard, Mechanism of four-electron reduction of dioxygen to water by ferrocene derivatives in the presence of perchloric acid in benzonitrile, catalyzed by cofacial dicobalt porphyrins, J. Am. Chem. Soc. 126 (2004) 10441.
[89] C.J. Chang, Z.H. Loh, C. Shi, F.C. Anson, D.G. Nocera, Targeted proton delivery in the catalyzed reduction of oxygen to water by bimetallic pacman porphyrins, J. Am. Chem. Soc. 126 (2004) 10013.
[90] (a) T.J. Stockmann, J.-M. Noel, A.A. Hassan, C. Combellas, F. Kanou, Facilitated lewis acid transfer by phospholipids at a (water|CHCl3) liquid|liquid interface toward biomimetic and energy applications, J. Phys. Chem. C 120 (2016) 11977.
(b) S. O'Sullivan, E. Alvarez de Eulate, Y.H. Yuen, E. Helmerhorst, D.W.M. Arrigan, Analyst 138 (2013) 6192.
(c) M.L. Colombo, J.V. Sweedler, M. Shen, Nanopipet-based liquid–liquid interface probes for the electrochemical detection of acetylcholine, tryptamine, and serotonin via ionic transfer, Analytic. Chem. 87 (2015) 5095.
(d) R.A. Hartvig, M. van de Weert, J. Østergaard, J. Jorgensen, H. Jensen, Formation of dielectric layers and charge regulation in protein adsorption at biomimetic interfaces, Langmuir 28 (2012) 1804.
[91] H.-F. Wang, L.-H. Yu, X.-B. Wang, R.-M. Ho, A facile method to fabricate double gyroid as a polymer template for nanohybrids, Macromolecules 47 (2014) 7993.
[92] S.-M. Kang, A. Kumar, C.-H. Choi, K.E. Tettey, C.-S. Lee, D. Lee, et al., Triblock cylinders at fluid–fluid interfaces, Langmuir 30 (2014) 13199.
[93] J.A. Dolan, B.D. Wilts, S. Vignolini, J.J. Baumberg, U. Steiner, T.D. Wilkinson, Optical properties of gyroid structured materials: from photonic crystals to metamaterials, Adv. Optical Mater. 3 (2015) 12.
[94] H. Karcher, The triply periodic minimal surfaces of alan schoen and their constant mean curvature companions, Manuscripta Mathematica 64 (1989) 291.
[95] S. Vignolini, P.J. Rudall, A.V. Rowland, A. Reed, E. Moyroud, R.B. Faden, et al., Pointillist structural color in Pollia fruit, Proc. Natl Acad. Sci. U.S.A. 109 (2012) 15712.
[96] K. Michielsen, D.G. Stavenga, Gyroid cuticular structures in butterfly wing scales: biological photonic crystals, J. R. Soc. Interface 5 (2008) 85.
[97] K.A. Asghar, D.A. Rowlands, J.M. Elliott, A.M. Squires, Predicting sizes of hexagonal and gyroid metal nanostructures from liquid crystal templating, ACS Nano 9 (2015) 10970.

[98] J. Allgaier, A. Poppe, L. Willner, D. Richter, Synthesis and characterization of poly [1,4-isoprene-b-(ethylene oxide)] and poly[ethylene-co-propylene-b-(ethylene oxide)] block copolymers, Macromolecules 30 (1997) 1582.
[99] S.C. Warren, F.J. DiSalvo, U. Wiesner, Nanoparticle-tuned assembly and disassembly of mesostructured silica hybrids, Nat. Mater. 6 (2007) 156.
[100] S.W. Robbins, P.A. Beaucage, H. Sai, K.W. Tan, J.G. Werner, J.P. Sethna, et al., Block copolymer self-assembly-directed synthesis of mesoporous gyroidal superconductors, Sci. Adv. 2 (2016) e1501119.
[101] (a) X.-C. Luu, J. Yu, A. Striolo, Ellipsoidal Janus nanoparticles adsorbed at the water–oil interface: some evidence of emergent behavior, J. Phys. Chem. B 117 (2013) 13922.
(b) B.J. Park, C.-H. Choi, S.-M. Kang, K.E. Tettey, C.-S. Lee, D. Lee, Double hydrophilic Janus cylinders at an air–water interface, Langmuir 29 (2013) 1841.
[102] T. Zhang, H. Qi, Z. Liao, Y.D. Horev, L.A. Panes-Ruiz, P. Petkov St., et al., Nat. Comm 10 (2019) 4225.

CHAPTER 4

Synthesis and generation of polymeric materials from interfaces

4.1 Introduction

Polymer science and technology have had an enormous impact on society and industry since the advent of the first polymeric material by Charles Goodyear in 1830. This landmark was achieved through vulcanization, a process that induces crosslinks among polyisoprene chains. In 1847 the reaction of cellulose with nitric acid was shown to lead to the formation of a celluloid material, whereas more polymeric materials, such as bakelite, could be synthesized through a condensation reaction. Other important examples include the perfluorinated heat-resistant polymers such as Teflon, Nomex, and Kapton. In 1938 Roy Plunkett, who was working as a research chemist in the factory plants of the DuPont Chemical company, accidentally synthesized and discovered the polytetrafluoroethylene, a polymer of tetrafluoroethylene that showed excellent resistance to corrosion and heat. This is the commonly known Teflon with many everyday life applications.

We can all state examples of polymer use in everyday life. All major chemical industries are competing for the discovery of new materials and the improvement of the properties of existing ones. For example, in 2000 the annual production of polymers in the United States reached 87 billion lbs, with thermoplastic polymers such as polyethylene (PE), polypropylene (PP), polystyrene (PS), and Nylon representing most of the polymers produced. Fig. 4.1 shows the chemical structure of various polymers that have found numerous applications based on societal needs.

Polymers are among the most widely used and applied materials and have reshaped the daily life of millions of people. Plastic materials have found applications as elastomers due to their high extension. Synthetic nitrile and butyl rubber polymers can imitate natural rubber. Nylon and polyester are excellent alternatives for natural fibers (i.e., cotton, wool, silk). The research and development departments of DuPont have

Generation of Polymers and Nanomaterials at Liquid-Liquid Interfaces
DOI: https://doi.org/10.1016/B978-0-12-819491-1.00004-6

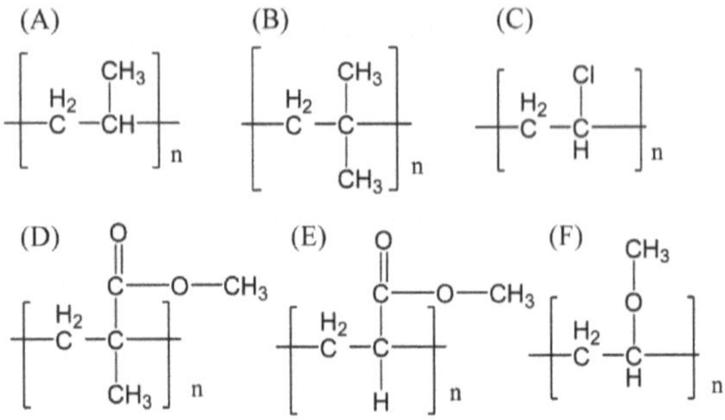

Figure 4.1 The chemical structure of a series of conventional polymers. (A) polypropylene; (B) polyisobutylene; (C) polyvinyl chloride; (D) polymethyl methacrylate; (E) polyvinyl acetate; and (F) polyvinyl methyl ether.

pioneered the synthesis and large-scale production of thermal- and flame-resistant polymers such as Teflon and Nomex, which have been used widely in harsh environments and specialized applications. Nomex is produced by condensation reaction from the monomers *m*-phenylenediamine and isophthaloyl chloride.

However, all these plastic polymers were hard insulators in terms of electrical resistance and exhibited very low conductivity values. The picture changed with the discovery of electronic transport properties in bulk polyacetylene. This started taking shape toward the end of 1970s, the advent of conductive polymers and their "tunable" and unique electronic properties, paved the way for a range of opportunities and potential applications for polymeric materials and chemical industry. Due to their sufficiently high and tunable electrical conductivity, these materials can be used as sensors or nanoelectronic components and as electrodes for energy storage. Furthermore perfluorinated polymers may be used as carbon dioxide (CO_2) thickeners for enhanced oil recovery and as superhydrophobic coating.

In this chapter the synthesis of conductive or photo-functional polymeric materials with improved crystallinity and preferable oxidation-states through liquid–liquid interfaces and their composites with silver or carbon nanomaterials are emphasized. Interfacial polymerization is a reaction that culminates in improved electronic transport and tuning of the fluorescence of newly developed oligomers and polymers. Thin films synthesized through interfacial polymerizations have also been proven through simulation studies to have a dense core and looser ends, leading to a heterogeneous distribution of functional groups and charge across the film.

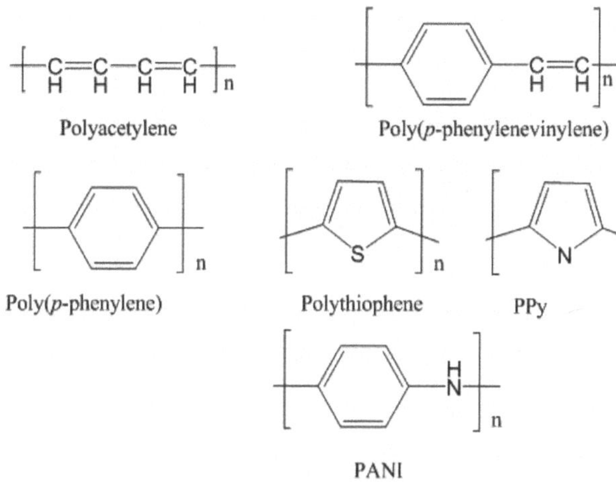

Figure 4.2 The chemical structure of the most important and widely used conjugated and potentially conductive, under certain doping levels, polymers. PANI is the abbreviation for polyaniline, and also polyacetylene, poly(*p*-phynylenevinylene), poly (*p*-phenylene), polythiophene, and polypyrrole are presented.

The chemical structure of the most important conductive and conjugated polymers is shown in Fig. 4.2. This includes polyacetylene, poly(*p*-phenylenevinylene), poly(*p*-phenylene), polythiophene, polypyrrole (PPy), and polyaniline (PANI). In this chapter, the discussion is dedicated to novel conductive polymers, conventional polymers, and composites, and in Chapter 5, Light Emitting Materials Generated at the Liquid-Liquid Interface, to light-emitting oligomers and two-dimensional (2D) materials with exciting optical and photophysical properties. These metal–organic polymeric materials may be synthesized through liquid–liquid interface systems and expand the gallery of 2D materials beyond those synthesized through chemical vapor deposition. This has been proven to be an increasingly important class of materials, especially since the groundbreaking physics that the advent of graphene and other 2D materials such as boron nitride (BN), chalcogenides, and black phosphorus have revealed.

4.2 Plastic polymers synthesized through addition or condensation polymerizations

In the present and Chapter 5, Light Emitting Materials Generated at the Liquid-Liquid Interface, we focus the discussion primarily on newly developed photoresponsive, metal–organic polymers and conductive or light-emitting polymeric nanomaterials. However, the interfacial polymerizations were introduced first for plastic polymers and produced relatively high molecular weight products such as polyamides (PA), polyesters, and polyurethanes (PU). Regarding the conventional polymers, the interfacial polymerization had been developed decades earlier as a methodology for obtaining nylon films [1]. Initial efforts to produce polyamides through the reaction of selected diamines with diacid chloride in aqueous environments resulted in products with low molecular weight. Furthermore the aqueous medium was shown to be ineffective for the accomplishment of the reaction [2]. To that end, polyurethanes have been synthesized by reacting bischloroformates with diamines in an interfacial system, with hexamethylenediamine and tetramethylene *bis*(chloroformate) as the molecular precursors. The synthetic strategy through liquid–liquid interfaces led to a narrower distribution of the molecular weight. Fig. 4.3 shows the reaction scheme of sebacoyl chloride dissolved in benzene, which reacts with hexamethylenediamine (which is dissolved in water). The bridging amide groups, –NH–CO–, in the nylon chains exist almost exclusively in the trans-conformation, and their H-bonding renders them substantially stiffer than, for example, $-CH_2-CH_2-$ groups. In the case of nylon the polymer can be readily removed from the interface as a film.

Other examples are the bulk polymer synthesis of polycarbonates (PC) [3] and the synthesis of microcapsules that can "trap" enzymes [4]. In the work by Chang and coworkers, interfacial polymerization was used in a complex three-solvent system comprised of water–chloroform–cyclohexane in the presence of selected detergents. The microcapsule size was found to be

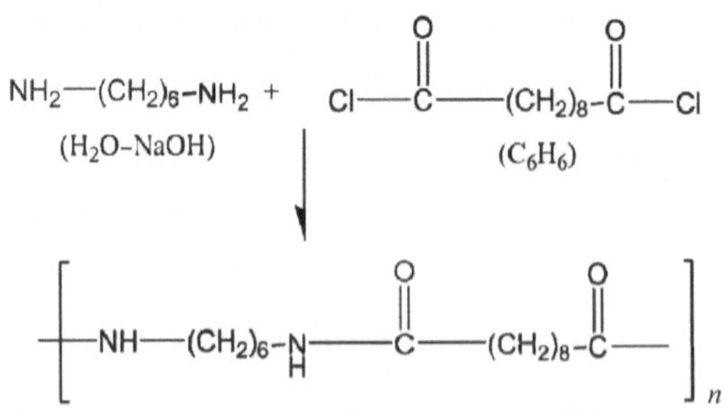

Figure 4.3 The interfacial polymerization of a polyamide in a water–benzene system leads to polyamine thin films that can be removed easily from the interface.

dependent on the concentration of the surfactant added in the mixture. The surfactant could be Span 85, Tween 20, or *n*-butyl benzoate. Protein addition in the aqueous phase at high concentration facilitated capsule formation. Other works have shown the synthesis of thin-film composite membranes [5] and PA—carbon nanotube (CNT) nanocomposites for reverse osmosis. Other polymer nanocomposites through interfaces have been reported by Moniruzzaman and coworkers [6]. They synthesized polymer—nanotube nanocomposites by employing surface-functionalized SWCNT in an interfacial polycondensation method. CNTs were decorated with $-(CH_2)_nCOCl$ functional groups on the sidewalls, which enabled them to be cross-linked to a Nylon matrix via alkyl segments. The Nylon—nanotubes composites were shown to possess a superior tensile modulus, strength, and toughness compared with pristine Nylon and Nylon composites modified with nonfunctionalized SWCNT. These properties are shown in Fig. 4.4 and were assigned to the alkyl linkages on the CNT—Nylon 6,10. These linkages were found to reduce the flexibility of the Nylon matrix, which explained the difference that was observed in mechanical properties. The experiments demonstrated that with a 2% wt addition of CNTs in the polymer matrix, a 214% increase on the elastic modulus, and an 162% increase on the yield strength can be achieved. The authors identified dichlorobenzene as an optimum organic solvent. The organic-to-aqueous-solvent ratio (1.2, giving an 80% reaction yield) and optimum diamine concentration (10% excess) were also identified. Initially the MWNTs are dispersed in dichlorobenzene and sebacoyl chloride

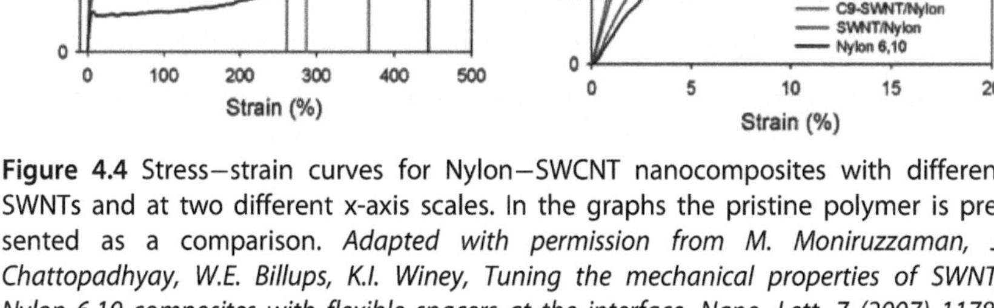

Figure 4.4 Stress—strain curves for Nylon—SWCNT nanocomposites with different SWNTs and at two different x-axis scales. In the graphs the pristine polymer is presented as a comparison. *Adapted with permission from M. Moniruzzaman, J. Chattopadhyay, W.E. Billups, K.I. Winey, Tuning the mechanical properties of SWNT/Nylon 6,10 composites with flexible spacers at the interface, Nano. Lett. 7 (2007) 1178., copyright American Chemical Society.*

is subsequently added to the suspension. It was also found that the addition of functionalized SWNT slightly reduced the molecular weight of the resulting nylon (the obtained molecular weight was estimated to be 16,000 g/mol in 1 wt.% composites). The polymer crystallizes anisotropically as proven by X-ray diffraction (XRD) experiments.

Haggenmueller et al. presented the in situ polymerization synthesis for nylon 6,6 and composites with single-wall carbon nanotubes (SWNT). The CNTs were used in purified, functionalized, or surfactant stabilized forms and they were characterized by FTIR, Raman spectroscopy, and thermogravimetric analysis (TGA). A large variation in modulus for the purified SWNT–nylon 6,6 composite fibers with 2 wt.% was observed and is associated with the inhomogeneous CNTs dispersion (Fig. 4.5) [7].

The decrease of the resistivity in nanocomposites where conductive nanomaterials have been incorporated as fillers is a complicated phenomenon and depends not only on the concentration but also on the dispersion

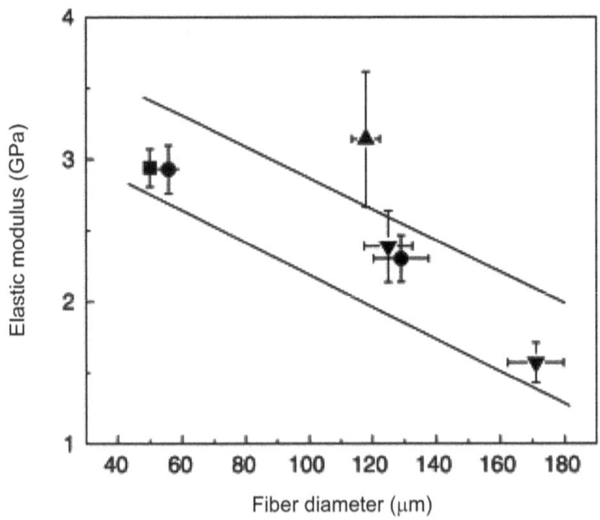

Figure 4.5 This graph represents the elastic modulus values obtained from tensile tests on individual melt-spun fibers. The deformation rate was 10 mm/min. The samples and the respective markings that are presented are: (■) nylon 6,6, (▲) 2 wt.% purified SWNT/nylon 6,6, (▼) 2 wt.% SWNT–f12/nylon 6,6, and (●) 1.6 wt.% SWNT–f18/nylon 6,6. *Reproduced with permission from R. Haggenmueller, F. Du, J.E. Fischer, K.I. Winey, Interfacial in situ polymerization of single wall carbon nanotube/nylon 6,6 nanocomposites, Polymer 47 (2006) 2381–2388, © Elsevier.*

and the interaction between the filler and the polymer matrix [7]. An example is the case of CNTs that are not oriented and well dispersed. According to simple percolation models, increasing the aspect ratio results in a percolation threshold at lower carbon loadings. The agglomeration of the fillers depends on many factors including the low shear forces, surface functionalization, and interactions with the matrix (Fig. 4.5). This has a tremendous influence on both the electrical conductivity and mechanical properties. Fig. 4.5 shows the elastic moduli of nylon−carbon fiber composites as a relationship with the diameters of the fibers. The diameter of the material ranged from 40 to 170 μm.

In the work by Ichiura et al. [8] the aim was to control the synthetic and fabrication procedures of a novel functional paper. Their methodology was based on polyamide (nylon-2,8) that was prepared on the surface of the paper by employing interfacial polymerization. For the formation of nanocomposites, the filter paper was impregnated with ethylenediamine (EDA) solution and NaOH solution and was placed in a beaker containing an organic solvent. Examples of organic solvents that were employed are cyclohexane, chloroform, or a cyclohexane − chloroform solvent mixture (volume ratio = 3:1, 1:1, 1:3), or a mixture of terephthaloyl chloride and trimesoyl chloride. The morphology of the nylon-2,8 polymer changed according to the concentration of ethylenediamine and the organic solvent that was employed. When mixtures of cyclohexane − chloroform were used, the morphology depended to a large extent on the EDA concentration due to the interfacial phenomena that are taking place at the paper surface. These phenomena are represented by the interfacial tension between the aqueous and organic phase and the dispersion of the ethylenediamine from the aqueous to the organic phase. The authors found that cyclohexane lead to the formation of a nylon-2,8 capsule-like structure on the paper surface. For cyclohexane − chloroform as solvents, the nylon-2,8 morphologies were fibrous or porous according to the EDA concentration and the organic solvent that was used. This was assigned to two main differences. First in the spreading coefficient (S) of the EDA solution adsorbed on the paper surface and second on the partition coefficient (P) shown by the dispersion of EDA fixed on the paper from the aqueous to the organic phase. The absorption of acetaldehyde was also dependent on the morphology of the nylon polymer and the optimum conditions were found to be 10% EDA and cyclohexane:chloroform solvent ratio of 3:1 [9].

4.3 Oxidative polymerization of conductive polymers

4.3.1 Polyaniline: interfacial polymerization of aniline

4.3.1.1 Principles and formation mechanism of nanofibres at the interface

Conductive polymers have had an enormous impact on materials science. This is because until the discovery of conductive polyacetylene, all the known and studied polymers were insulators with extremely low conductivity values. The discovery of electronic transport in conductive polymers gave a new direction to materials science, engineering, and physical chemistry, leading to the award of the Nobel Prize in Chemistry in 2000 to the three scientists who pioneered this field: Hideki Sirakawa, Alan Heeger, and Alan MacDiarmid. Among the series of conductive polymers developed since the late 1970s, PANIs perhaps remain the most studied member of this family of materials. This is due to its high conductivity; ability to form readily from oxidation and protonation -states, with each one exhibiting distinct optical and electronic properties; light weight; low cost; and excellent synthetic yields [10]. On the other hand PPy suffers from low solubility, and polyacetylene is not particularly stable in air.

Due to its linear and pi-conjugated backbone, PANI is one of the most well-known polymers that can form one-dimensional structures such as "nanowires," "nanoneedles," and "nanofibers." While CNTs and carbon fibers along with other linear materials are well known and have been repeatedly studied, this was of particular impact for PANI. The linear morphology has a crucial role in the properties of the conductive polymers, and these linear, 1D, polymer nanomaterials are particularly useful in sensor applications. This behavior is assigned to the very high specific area of the fibers, which enables rapid diffusion, reaction, and so detection of small molecules that can cause immediate changes in the oxidation and protonation states and elicit immediate effects on electronic transport and optical properties. For the synthesis of this type of fibrillar polymeric material, interfacial polymerization has been proposed to be perhaps the most efficient methodology.

Initially this chapter discusses the oxidative polymerization mechanism of aniline since it is the most abundant and readily synthesized nanofibrillar polymeric nanomaterial that possesses conductive properties (Fig. 4.6). Usually this process involves the addition of an acid and an oxidant in the aqueous phase and a monomer in the organic phase. Usually the oxidant are persulfate salts. In this organic—aqueous interfacial system, the aniline

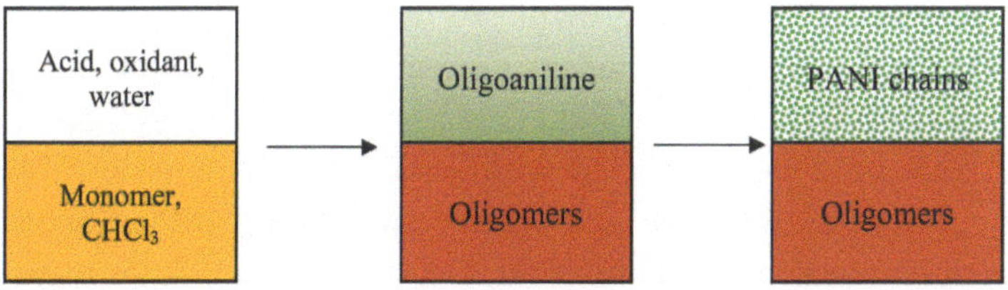

Figure 4.6 The steps taking place during the interfacial polymerization of aniline at liquid–liquid interface (schematic representation with an organic solvent that is heavier than water, for example, chloroform or dichloromethane). The monomer is added to the organic phase and the oxidant (in most cases, the oxidant is persulfate ions, $S_2O_8^{2-}$) and the acids are added in the aqueous phase [3]

monomer tends to be oriented with the polar amine group heading toward water (where oxidation is initiated) and the nonpolar group headed toward the organic phase. The liquid–liquid boundary is restricted space, hence the first oligomers are formed in the 2D space of the interface, which is also suppressing secondary growth of the polymer. To that end elongated anisotropic structures are formed.

To identify the mechanism of interfacial polymerization, many research works have focused on the isolation and subsequent study of the oligomers and polymers formed in the various stages during the polymerization and the chain growth of aniline. The typical electrochemical oxidation states of PANI as obtained in a cyclic voltammetry experiment can be seen in the work by Huang and Kaner [11]. If we consider the interfacial polymerization compared with a conventional one-phase synthesis, in the former case there are fewer nucleation sites that propagate the polymer reaction. In single-phase polymerization, initiators are mixed with the monomer in a 3D space. Hence each oxidized monomer can create a nucleation center for polymer chain growth. In contrast if the process takes place at the organic–aqueous interface, the interaction between initiator molecules and monomers is essentially limited to the restricted 2D space.

The isolation, purification, and study of oligomers and other intermediates mentioned above have provided strong evidence that the fibrillar morphology of PANI is an intrinsic characteristic of the polymer, whereas the irregular and globular shapes that are observed routinely in one-phase systems are formed in the "secondary growth" of macromolecules [12]. This secondary growth may be considered as undesired based on the previously mentioned reasons.

In general the polymer growth is a multistep procedure: the initial nucleation step and the secondary growth step are separated. Thus the synthesis of nanofibres, nanotubes, nanocomposites, nanoneedles, and nanospheres must be controlled with sufficiently good yields. This secondary growth is the basic idea behind the efficient separation through interfacial polymerization. During the formation of 1D nanostructure, the secondary growth of polymer chains can be suppressed. Consequently the overall and relative yield of nanofibres in the final product can be increased substantially, leading to the preferential synthesis of elongated and anisotropically shaped structures due to the minimization of secondary growth. The monomer (e.g., aniline, pyrrole, 3-ethylene dioxythiophene) and the initiator (e.g., ammonium peroxydisulfate, iron nitrate) are separated through the boundary between the aqueous and organic phase, so the polymerization occurs at the interface. Due to the absence of any stirring, all the components that are necessary for the polymerization are forced to diffuse in this restricted interfacial space [3b] and subsequently the polymer grows in elongated nanostructures [13].

During the polymerization various structures are formed with differences in the molecular weight(s) (MW), doping level, and oxidation state. In general the nanofibres are readily formed in the doped, conductive, emeraldine salt form of the PANI backbone. They are slightly hydrophilic, so they can diffuse into the water phase. The interfacial reaction enables the as-formed fibers to avoid unwanted secondary growth. This allows new nanofibres to grow continuously at the liquid—liquid interface independently irrespective of the organic solvent used (i.e., chloroform, dichloromethane, toluene). Aniline monomers can also be polymerized in the confined space of the monolayers of lipid membranes in the presence of strong inorganic and organic acids. Mass diffusion still leads to the formation of the anilinium cation during the first stages of polymerization. This aromatic cation, as an intermediate step of polymerization, can be adsorbed strongly in the charged monolayers of the phospholipids L-a-dimyristoyl-phosphatidic acid and hexadecanesulfonic acid at the air—aqueous interface and subsequently leads to a chemical polymerization with organic surfactants acting as soft reaction templates [14].

4.3.1.2 Nonfibrillar morphologies that can be generated at an Interface

There is a wealth of morphologies that may be generated during this interfacial procedure, for example, by altering the concentration of a surfactant that is added to the aqueous phase [9,10]. When the commonly used cetyltrimethylammonium bromide (CTAB) was employed during oxidative polymerization of aniline, various morphologies, from needle-shaped (at a CTAB concentration of 0 mol/L), rice-like (0.02 mol/L), hollow microspheres (0.04 mol/L) to wires (0.06 mol/L) were observed [15]. CTAB is a well-studied amphiphilic molecule with a critical micelle concentration in an aqueous solution of ≈ 0.92 mM [16].

Li and coworkers studied different methods of interfacial and micellar polymerization of aniline in the presence of strong inorganic acids such as hydrochloric acid. They used CTAB and undertook polymerization reactions with different concentrations of the surfactant. They found that the morphology of hydrochloric acid-doped PANI could be tuned from 1D nanoneedles or nanowires that formed a network (with a diameter approximately 50–100 nm) to 3D hollow microspheres (with a diameter of 400 nm).

As discussed in previous chapters, a surfactant arranges itself in an aqueous solution in a specific way: polar groups are headed toward the aqueous phase and aliphatic chain heads are toward themselves to avoid water so that they can act as reactions templates. During the polymerization of aniline, these micellar templates suppress the tendency of the polymer to grow in 1D nanostructures, and isotropic (or even hollow) structures can be synthesized. The pioneering work by Huang and coworkers demonstrated that interfacial polymerization is an effective and extremely efficient method used to suppress the secondary growth of PANI. This is achieved by the formation of nanofibres and nanoneedles in the restricted 2D space of the liquid–liquid interface. This strategy is in contrast with a conventional homogeneous reaction system and, according to Li and coworkers, a higher CTAB concentration lead to a change in the morphology of micelles from spherical to linear. In other works nanostructured PANI was synthesized with polyacrylic acid to act as a counterion [17]. Ding and coworkers synthesized sodium dodecyl benzenesulfonate-doped PANI nanorods through interfacial polymerization and, according to their ultraviolet (UV)–visible spectra, the PANI seemed to be in the nondoped, insulating form [18].

Hence the interfacial reaction of aniline may lead to a wealth of nanostructures and, depending on the density of the organic medium, may be controlled through variation of the polarity of the interface; interfacial tension; addition of various surfactants or dopants; solubility of the products; and gravitational factors. Further applications of the system for pyrrole polymerization, composite materials, and light-emitting polymers and oligomers are presented in the following section.

4.4 Interfacial polymerization of pyrrole: formation of thin films and noble metal nanocomposites

PANI is a polymer with a slightly hydrophilic character. It can be solubilized in water upon high dilution and sonication in order to stop unwanted clustering. In contrast with PANI PPy is an insoluble polymer that cannot be solubilized in any polar or nonpolar solvent (organic or water). Based on this unique lack of solubility for the PPy black, the polymerization in a liquid—liquid interface should be expected to lead to the formation of a suspended thin film at the boundary of the two phases, which can be removed from the interface and purified with reaction solvents readily. The lack of solubility of the black PPy powder, in contrast with its oligomers, in any solvent has been assigned to extensive crosslinking between polymer chains [19]. For a detailed analysis of the mechanism of this interfacial polymerization, see the review by Dallas and Georgakilas [20]. Fig. 4.7 shows the mechanism for oxidative formation of PPy

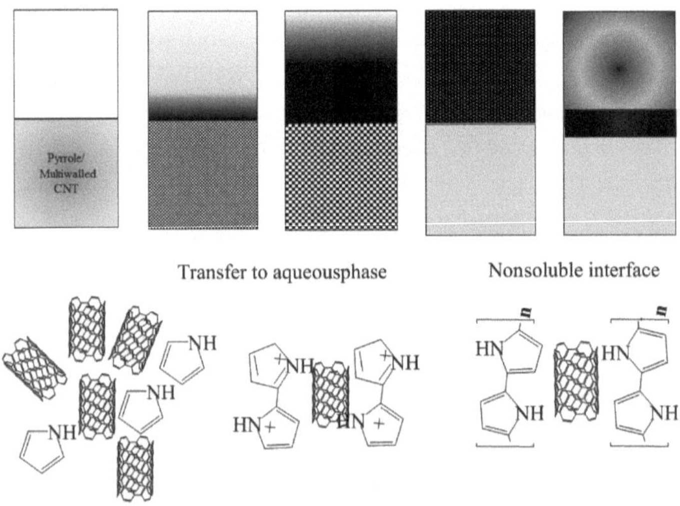

Figure 4.7 The interfacial oxidative polymerization of pyrrole toward the synthesis of carbon nanotubes-PPy nanocomposites. *Reproduced with permission from P. Dallas, V. Georgakilas, Interfacial polymerization of conductive polymers: generation of polymeric nanostructures in a 2-D space, Adv. Colloid Interf. Sci. 224 (2015) 46, adapted with permission Elsevier.*

through a redox reaction with metal cations. It involves oxidation of the monomer with metal cations, which could be ferric ions derived from iron nitrate or silver cations derived from silver nitrate (for Fe^{3+}; $E_{ox} = 0.77$ V) and (for Ag^{+}; $E_{ox} = 0.80$ V), both of which are dissolved in aqueous solutions. The first report of pyrrole polymerization was an oxidative polymerization occurring in aqueous solutions, where the addition of iron nitrate led to the spontaneous formation of black PPy powder. Usually the reaction takes place in aqueous solutions in the presence of an oxidizer. The pyrrole monomer has a solubility of 7.5 wt.% in water [21]. The most commonly used oxidative agents are ferric ions, which oxidize the monomer by attacking the α-position of the pyrrole ring, and the reaction propagates through the subsequent addition of pyrrole molecules. The black PPy precipitate that is instantly formed is a charged polycations with cross-linked chains, nitrate as a counterbalancing anion and it famously lacks solublility in any polar or nonpolar solvent.

Cyclodextrins are highly soluble macromolecules in the water thanks to their decoration with hydroxyl groups. Their remarkable solubility in water has made them appealing candidates for the synthesis of inclusion complexes and biomedical applications. They have also been applied in conductive polymers and for the synthesis of a 3D network from branched PPy nanofibres through interfacial polymerization. The interfacial reaction takes place in the presence of the sulfonated β-cyclodextrins (Sβ-CD) which, due to the sulfonate ions, act as a doping agent. The effect of the concentration of Sβ-CD in polymerization solvents, the type of oxidant used to initiate the reaction, and the temperature on the electrical properties of PPy−Sβ-CD was investigated [22]. Similarly interfacial polymerization in the presence of Sβ-CD has been reported for the aniline monomer [23] and led to the formation of chiral PANI nanobundles. The chiral nanostructures stem from the intertwining of PANI and cyclodextrin sulfate chains during polymer growth and they were isolated exclusively from aqueous solution irrespective of the organic solvent used.

4.5 Conductive polymer nanocomposites

4.5.1 Noble metal nanocomposites synthesized through reduction of metal ions by monomers

Pyrrole is also polymerized through oxidative or electrochemical polymerization similar to PANI. The most commonly used initiators are ferric ions and they are routinely used for the oxidation and subsequent polymerization of pyrrole. Similarly noble metal cations could have suitable oxidation potential and act as initiators. To that end, when silver, gold, or platinum cations are used as initiators, they are reduced and oxidize pyrrole monomers simultaneously. This strategy initiates polymerization and also forms noble metal nanoparticles (NPs). These noble metal NPs are immediately trapped from the nitrogen-based polymer chains that are growing in parallel with the inorganic NPs. This one-step polymerization is a versatile reaction and has been applied repeatedly for the formation of small noble metal NP dispersed finely within a conductive polymer matrix with a narrow size distribution [24]. For a typical reaction between pyrrole and silver nitrate as an oxidizer, 0.5 mL (7 mmol) pyrrole and 180 mg $AgNO_3$ (1.05 mmol) lead to the formation of a silver nanocomposite. For this molar ratio between silver and pyrrole we obtained a composite with a 60 wt.% silver content. This was estimated through thermogravimetric analysis, and transmission electron microscopy demonstrated formation of silver NP with a diameter of 5–10 nm. This method can be expanded readily to other noble metals such as gold, platinum, or rhodium. Similarly, PANI–silver nanocomposites have been prepared by employing an interfacial polymerization using ammonium persulfate as an oxidizing agent. Gold NPs of various sizes and shapes have been synthesized with poly(*o*-methoxy aniline) and gold chloride. This method resulted in the synthesis of spherically shaped or truncated pentagonal/hexagonal-shaped NPs dispersed within the polymer matrix [25]. Other works have focused instead on the dispersion of silica particles within polymer chains [26].

In a typical procedure Ag NPs can be prepared by reducing Ag salt with $NaBH_4$ a reducing agent in poly(vinyl alcohol) (PVA) solubilized in microemulsion of polyethylene glycol *p*-(1,1,3,3-tetramethylbutyl)-phenyl ether surfactant (also known as Triton X-100). Solution casting of Ag–PVA nanocomposite after separation from microemulsion gives Ag–PVA nanocomposite film with uniform and homogeneous arrangement of Ag NPs in the smooth PVA matrix. In ex-situ approach, it has been difficult to disperse Ag NPs homogeneously into a polymer matrix for the easy agglomeration of Ag NPs, especially at high concentrations [27]. On the other hand, for in situ formation of Ag–PVA using a reducing agent, the casted film might contain unreacted silver salts. The presence of these silver salts might cause toxicity through adsorption of ions in epidermis cells and sweat glands [28]. For in situ reduction and encapsulation of Ag NPs in PVA by the host itself, HNO_3 is produced as a byproduct at elevated temperatures [29]. This formation of acidic byproducts might restrict its use in dermatological applications. Again even in the presence of surfactants, the agglomeration still remains an issue at a high concentration of silver NPs. In this work, the composite film from in situ synthesis using a reducing agent, gave spherical bunches of aggregated Ag NPs that exhibited pits and cracks on the PVA film surface. The Ag–PVA nanocomposite film prepared using microemulsion exhibited remarkable enhancement of antibacterial activity against *Escherichia coli* and *Staphylococcus aureus* as compared with that prepared from bulk synthesis. The bacterial inhibitory property of the film prepared from microemulsion was also significantly higher compared with literature values [30]. The template role of microemulsion in combination with the PVA nanoreactor provides homogeneous and uniform distribution of Ag NPs which ensures the controlled release of NPs. Thus the as-synthesized film might be a potent bactericide to reduce and eliminate the risk of infections and be used for treatment of a patient simultaneously infected by Gram-negative and Gram-positive bacteria. This will be a step forward for the development of an efficient means for the fabrication of nanocomposite film with desirable size and distribution of NPs using microemulsions as a nanoreactor for enhanced concurrent antibacterial activity against Gram-negative and Gram-positive bacteria for biomedical applications.

4.5.2 Carbon nanotubes and graphene nanocomposites

The surface functionalization of CNTs through noncovalent bonds can be achieved through amino-containing pyrene derivatives with various spacer chain lengths. The amine with the longest spacer length (e.g., 12 methylene units) is the one that demonstrated the highest functionalization efficiency for SWCNTs in terms of dispersibility with 10 wt.% of the amine strongly adsorbed on SWCNTs through $\pi-\pi$ interactions [31]. Poly(*n*-butyl methacrylate) can also be grafted onto multiwalled CNTs using atom transfer radical polymerization. The enhancement of the mechanical performance of poly (vinyl chloride) using poly (*n*-butyl methacrylate)-*g*-MWNTs was studied and the miscibility between poly(vinyl chloride) and poly(*n*-butyl methacrylate) enabled the homogeneous dispersion of nanotubes in the matrix. At 0.2 wt.% CNTs content, the addition of poly (*n*-butyl methacrylate)-g-MWNTs increases in storage modulus (at a temperature of 35°C), Young's modulus, yield stress, tensile strength, ultimate strain, and toughness of PVC by 83%, 40%, 74%, 84%, 38%, and 145%, respectively, and the reinforcement values on Young's modulus and tensile strength are 387 and 19.1 GPa [32a].

Carbon nanomaterials have been at the forefront of materials science, research, and technology since the discovery of fullerenes in the 1980s and the later advent of CNTs and graphene. The discovery of fullerenes and the exfoliation of graphite toward single-layer graphene resulted in the award of two Nobel Prizes [33]. Incorporation of such carbon nanomaterials within polymer matrices, be it conventional or conductive polymers, would be of particular interest because CNT have been some of the stiffest and strongest man-made materials known to date, and graphene has provided exciting new opportunities for materials science [34]. CNT also exhibits very high electrical conductivity, which is very important for applications in electronic devices and communications [35]. The interfacial polymerization has been shown to lead to efficient grafting of nanotubes with a polymer layer. For that purpose CNTs are added to the oil phase after their functionalization with organophilic hydrocarbon molecules such as oleylamine. During polymerization the CNTs are trapped by the slightly hydrophilic pyrrole oligomers and are transferred toward the aqueous phase. After a certain period of time the molecular weight of the PPy increases substantially and, due to extensive crosslinking, the polymer is now insoluble in any solvent. Thus the composites form an interfacial film [36].

4.6 Applications and physical properties

4.6.1 Sensor applications

PANI is the most widely used conductive polymer due to its good environmental and air stability; water affinity; slight hydrophilicity; bulk synthesis in high quantities; and spontaneous reactivity with various chemical species. These properties have led to the construction of very efficient sensors, gas analytes, and liquid analytes [37]. Many PANI sensors involve spin-coated films of uniform composition that can react readily with reducing or oxidizing agents, and protonating or deprotonating acids/bases.

All these classes of materials are expected to lead to spontaneous alterations on the electronic conductivity and the optical properties of the polymer, resulting in sensing calibration curves that enable calculation of the analyst concentration. The mathematical relationship that shows the response of resistive sensors is $\Delta R/R_0$, where R_0 and R are the electrical resistance before and after contact between the polymer and the agent that will be detected, respectively, and $\Delta R = R - R_0$. The protonated emeraldine salt form of PANI is conductive, so the emeraldine base formed may be used for acid detection based on an increase in conductivity after exposure to acids. The dominant theory regarding the electronic conductivity in these polymers involves the formation of polarons, to be later discussed in more detail. This mechanism involves the movement of charge carriers along macromolecules in a reversible process. Many studies have compared the performance of PPy and PANI nanostructures as sensors. As humidity sensors, PANI nanorods exhibit negative response values ($R < R_0$) at low humidity, whereas PPy nanorods of similar size and shape show a positive response ($R > R_0$) [38]. The sensitivity of a sensor is defined as the slope of the response versus the concentration curve and both polymers demonstrate linear behavior with sensitivity values of -0.364 for PANI and 0.168 for PPy.

This change in conductivity values is brought about by the formation of "polarons" ("radical cations") [39]. Polarons and bipolarons are charge carriers in emeraldine salt that move along the polymer chain upon application of electric field and are responsible for the electrical conductivity of the polymer, as evidenced by electron paramagnetic resonance studies [40]. All these acid–base reactions are completely reversible (which is essential for a chemical sensor). I am focusing my discussion on the emeraldine salt (which is the conductive form) and its insulating, base form, but PANI can exist in several oxidation states. These oxidation states are the fully reduced leucoemeraldine form, the half-oxidized emeraldine form, and the fully oxidized pernigraniline form. These oxidation states can be controlled by altering and tuning the synthetic conditions or from postsynthesis oxidation–reduction reactions. For example, hydrazine can alter PANI from the emeraldine form to the leucoemeraldine form and thus lead to the development of polymeric sensors for the detection of hydrazine [41].

Due to the high conductivity of the emeraldine salt of PANI, most studies have been devoted to polymer sensors for small acids and bases. In a typical procedure the polymer is exposed to basic molecules such as ammonia gas [42]. Less research activity has been focused on use of the insulating emeraldine base form of PANI, which is used as a sensing material for acidic gases such as hydrochloric acid. Detection of acid is required in many fields and applications; such as rocket launches. In the latter hydrochloric acid is formed in the exhaust plumes of rocket motors after decomposition of ammonium perchlorate (which is used as rocket propellant).

This interfacial polymerization as a selective synthesis of nanofibers was initiated and pioneered by the Kaner research team. They were the first to develop PANI nanofibres that can be used as sensor materials for the detection of acids, bases, reducing agents, organic vapors, and alcohols [40,43]. To demonstrate the importance of fibrillar morphologies, thin films of nanofibres and standard PANI were compared. This comparison was used to evaluate the response time and sensitivity of conventional PANI with those of nanofibres in the same oxidation state. Electronic transport and electrical resistance were monitored as a function of time while the polymers were exposed to various gases. Examples of the analytes used were hydrochloric acid, ammonia, hydrazine, chloroform, and methanol, and each analyte induced a distinctive response on polymer contact: protonation—deprotonation, reduction, swelling, and conformational alignment, respectively. According to their work, the importance of a fibrillar morphology was demonstrated through the response time toward vapors of hydrochloric acid or ammonia. The work showed that the response was significantly better and optimized for thin films composed of PANI nanofibres compared with the response time observed for spherical conventional particles produced by a common one-phase polymerization process. The source of this improved, optimized, and selective response was assigned to the high surface area and porous structure. The porous structure is the key parameter that enables easy and rapid diffusion of gases through the film and they did not show any thickness (0.2—2.5 μm) dependence in their performance. This was proven after systematic studies on film thickness showed no difference in the response magnitude or time. The small diameter of nanofibres and their narrow size distribution leads to rapid diffusion of dopants within the polymer chains. All these factors explain why the nanofibres provide a faster and most significant response to dopant exposure compared with films. In order to clarify the influence of the doping/dedoping processes to the surface of the polymers, the authors performed Brunauer—Emmett—Teller (BET) surface area measurements. They found that the camphor sulfonic acid-doped nanofibers have a surface area of 41.2 m^2/g while after dedoping with base, the surface area increases to 49.3 m^2/g. So the volume of the nanofiber polymer increases after the dopants are removed. On the other hand, the surface area of the nanofibers increases when the average diameter decreases. The HCl dedoped nanofibers with a diameter of 30 nm is 54.6 m^2/g, while the CSA dedoped nanofibers with a diameter of 50 nm, 49.3 m^2/g or the $HClO_4$ dedoped nanofibers (120 nm, 37.2 m^2/g).

The ability of PANI as a colorimetric sensor for the detection of metal cations was recently demonstrated by Kim et al. Polyalinine NPs were synthesized by radical polymerization process using acidic solutions of hydrochloric and phytic acid, yielding chlorinated (Cl^-) and phytic acid-doped (Ph^-) PANI. It demonstrated a great affinity for copper ions, Cu^{2+}– and the adsorption efficiency was assigned to high capacity and/or high ionic affinity of the doped phytic acid [44]. An enhanced Cu^{2+} adsorption capability of PANI nanofibers (PANI NFs) was demonstrated by doping with phytic acid. The PANI doped with phytic acid showed a much higher copper adsorption efficiency than the PANI doped with chlorine anions. This was assigned to the high capacity and/or high ionic affinity of the phytic acid. The absorption of the copper cations was proven to be pH dependent with increasing aqueous pH and it was not affected by the presence of cobalt cation in the aqueous solution.

Polyaniline and its composites have been studied widely as sensors for ammonia detection because, through exposure to basic ammonia vapors, deprotonation occurs which leads to a decrease in conductivity [45]. More detailed studies on these systems revealed a relationship between the film thickness and the sensing response. As the film thickness is increased, the polymeric sensor appears to be less effective in terms of sensitivity and time of response [41]. Furthermore a larger amount of base is required for the complete conversion of the polymer. This is the reason for the slow response time in previous works on ammonia sensors. That is, fewer dopants are required to reach the percolation threshold for conductivity than those that must be removed to convert the polymer to the insulating form.

Any small amount of water would be detrimental because the product of the reaction between ammonia and hydrochloric acid (ammonium chloride) is not neutral. If ammonium chloride comes into contact with water, it will be hydrolyzed back to reform hydrochloric acid, which subsequently redopes the polymer chain, thereby diminishing the effectiveness of the sensor. This is the main reason that the response of a base sensor is slower than that of an acid sensor. Porosity, surface area, and film morphology explain why conventional polymers respond less efficiently to a dedoping through bases than nanofibres. Other niche composite materials can be used as sensors. Some recent conductive composite materials are based on the dispersion of PANI in cellulose paper for the synthesis of highly efficient sensors. The sensors work by tracking the pH changes of the paper upon exposure to acid or bases. The sheets of pressed cellulose fibers used by Souza and coworkers were obtained from *Eucalyptus grandis* and *Eucalyptus urophylla* [46]. These cellulose sheets were modified with PANI particles synthesized through chloroform and were subsequently dispersed and used as a colorimetric sensor [47].

Examples of other types of molecules that can be detected beyond acid–base detection and estimation of concentration are nitrogen oxides. It has been shown that the resistivity of a polymer increases constantly upon exposure to nitrogen oxide gas [29]. Regarding the selectivity of the whole process, PPy nanofibres have shown an excellent response to ammonia but showed no response to nitrogen oxides even at ≤ 100 ppm, which is a crucial selectivity difference compared with PANI [48a].

Polymeric switches have been synthesized through single-crystalline PEDOT needles from interfacial polymerization [48b]. The single crystalline PEDOT nanoneedles showed an abrupt switching near 3 V. This is due to the electron transport along the $\pi-\pi$ stacking direction of the PEDOT nanoneedles at the order of milliseconds.

Li et al. reported a hierarchical graphene oxide–PANI nanocomposite via liquid–liquid interfacial electrochemical polymerization and its application in supercapacitors [48c]. These composites have high specific capacitance, good rate capability, and cycling performance, hence they demonstrated a high energy density and maximum power density due to the p–p interactions between the graphene oxide and PANI. Furthermore, PPy –mesoporous carbon (denoted as OMC) composites were fabricated by Choi et al. via a liquid–solid interfacial polymerization method and they were applied for direct methanol fuel cell applications [48d].

4.6.2 Single crystalline conductive polymers: the ordering of polyaniline chains

As expected the ordering of the polymer chains will influence their physical and chemical properties. In general polymers are amorphous or semicrystalline. Their macromolecular chains have a range of MW and crosslinks and hence different behaviors during crystallization and solid-state packing. This is true for conventional polymer macromolecular chains for the most amorphous PS or PE. Conductive polymers can exhibit crystalline symmetries at $2\theta = 19$ and 25 degrees (the 25 degrees symmetry corresponds to a d-spacing of 3.5 Å, which is typical for pi-stacked rings), as observed in the XRD pattern and are assigned to two possible symmetries: one parallel and one perpendicular to the polymer chains [49]. Hence polymers are a combination of ordered crystalline domains surrounded by an amorphous area: the semicrystalline state [50].

As a landmark in the quest for crystalline polymers, Nuraje and coworkers synthesized single-crystalline nanoneedles of PANI, PPy, and poly (3,4-ethylenedioxythiophene) through interfacial polymerization [51]. The organic phase was chloroform and the polymers grew into highly ordered single crystalline NP with an elongated morphology. Polymer dimensions were kept relatively small (<100 nm). The authors undertook conductivity studies and the nanoneedles, according to resistivity measurements and *I*/*V* curves, exhibited conductance switching at a timescale of milliseconds.

Ma and coworkers developed another methodology for the creation of belt-shaped PANI NPs through interfacial polymerization. They doped the PANI backbone with dodecatungstosilic acid. The materials were characterized by Fourier-transform infrared spectroscopy, XRD, and scanning electron microscopy (SEM). Dodecatungstosilic acid was the doping agent and furthermore it acted as an adhesive force with water. The polymer doped with dodecatungstosilic acid had higher conductivity compared with the polymer doped with hydrochloric acid.

Park and coworkers used tartaric acid as the acid for initiation of the reaction and as a doping agent in a procedure that lasted <15 min to synthesize highly crystalline PANI. The XRD pattern was characteristic of an emeraldine salt type II (ES-II) doped PANI (reflections were observed at 2θ: 31.9, 34.1, 23.6, 52.58, with a corresponding d-spacing of 0.2083, 0.2627, 0.3767 and 0.1742 nm, respectively) [52,53]. The authors claimed a crystallinity of >95%, which is the highest reported in the literature (most report values <50%–65% [54]). This very high crystallinity is advantageous in terms of electronic transport and the response as sensors [55].

The growth mechanism of inorganic NP has been presented in many works. The major parameters that control it are the crystallization and precipitation kinetics, the Oswald ripening and the temperature of the reaction. Similar to inorganic NP, the crystallinity and morphology of polymeric NPs can be controlled by kinetic, thermodynamic and chemical means. Fast growth will lead to a disordered and random orientation of polymer chains and an amorphous particle. During the liquid–liquid interfacial polymerization, which is a slower procedure compared with the one-phase reaction, the orientation of polymer chains in a more ordered structure is favored [56].

4.7 Electronic and magnetic properties

4.7.1 Transport mechanism: conductivity of polymers

Flexible electronic materials are needed for many applications, but the highly insulating nature of organic materials (except multilayered graphite) prevented them from replacing metals as conductive elements. However, towards the end of the 1970s, it was discovered that the conductivity of polyacetylene can increase by a factor of 10^{12}. This colossal increase will enable the materials to reach the conductivity of copper (10^4 S/cm). This remarkable increase is achieved after doping with an electron acceptor (e.g., AsF_5) at a saturation level of 8%, with iodine, or with an electron donor such as an alkali metal ion [57].

Conductive polymers and oligomers, in general, are unstable as electron transport materials since as 1D semiconductors they suffer from an intrinsic Peierls instability induced by electron–lattice coupling [58]. The two main factors that influence the conductivity of these types of polymers are (1) the ability to transport charge carriers along the macromolecular aromatic backbone; and (2) whether these carriers are allowed to "hop" between adjacent polymer chains. Consequently intrachain and interchain interactions help to decide conductivity.

Fundamentally the electronic transport mechanism is dominated by electron hopping alongside polymer chains. Hence the conductivity depends on the ordering of the chains; the dimensions of the macromolecular network; the way the chains are interconnected; the doping level; and the type of doping agents [59]. It seems that specific doping ions, such as camphor sulfonic acid, appear to be the optimum choice, with the doping with them giving the highest conductivity values. The influence of crystallinity can be demonstrated through extensive pi–pi stacking of the aromatic units among adjacent macromolecular chains. This overlapping of pi orbitals leads to enhanced electron delocalization and charge transport. Extensive conjugation stabilizes the overlap of pi electrons and reduces the bandgap. Doping agents create holes (positive charge carriers) or add electrons near the lowest unoccupied molecular orbital levels [60].

The doping agents are organic or inorganic acids that act as counterbalancing anions in the PANI backbone on $-N=$ sites. Acid doping results in the formation of a polaron with low-energy transitions in the red-near infrared region of the spectrum, as demonstrated by the three-band UV–visible spectrum of conductive PANI [46b]. The solvent from which PANI is cast in thin films results in different chain conformations and consequently to different conductivities [61]. Surfactants seem to lead to higher crystallinity and more homogeneous morphology. The conductivity from an activation energy-limited process appears to lead to a more metallic type of conductance [62].

4.7.2 Polyaniline nanoparticles with metallic electronic transport

The conductivity of this class of materials may vary significantly depending on many factors. These factors arise mainly from their long-range ordering, molecular structure, oxidation, and protonation states. Their electronic transport properties in a wide temperature range may be metallic or semiconductive. Metals show an increase in conductivity whereas the temperature becomes lower. A semiconductor shows a decrease in conductivity at lower temperatures. Their magnetic properties will be discussed in the context of their electronic transport properties.

As 1D conductors these materials are intrinsically unstable and undergo metal-to-insulator transition as the temperature is lowered. Dallas and coworkers synthesized, through liquid–liquid water–chloroform interfacial polymerization, needle-shaped PANI with sufficiently good crystallinity. XRD patterns indicated a good overlap of phenyl rings and small dimensions with cross-linked NP. The electronic transport of these polymeric NP from 330K to 230K followed metallic behavior, and transition to an insulator occurred at this temperature. At 230K the maximum value of conductivity was 3.5 S/cm [63]. Resistivity values were plotted according to the Mott relationship $\ln(RT/R330)$ versus $1/T^{1/2}$, which is employed in solid-state physics for studying semiconductors over a wide temperature range [64]. According to curve fitting the material was considered to be a 1D conductor with an x value of ½ and a Mott temperature (which reflects the energy required for charge hopping) of 1.5×10^4K.

Lee and coworkers [65] presented, for the first time, a PANI that showed metallic behavior until 5K. This is the first report of an intrinsically metallic polymer. They synthesized the materials through a self-stabilized dispersion polymerization and measured a remarkable conductivity value of 1000 S/cm, with the resistivity decreasing until 5K. Reflectance and dielectric spectra gave an excellent fit with the Drude model (even at the lowest frequencies) with a plasma frequency of 10.5/cm (1.3 eV) and fitting with the Hagen–Rubens approximation. Finally the values for other organic nanomaterials, for example, SWCNT were a plasma frequency edge at 2000/cm and conductivity of $\approx$ 170 S/cm.

4.7.3 Magnetic properties of conductive polymers

In Section 4.7, the electronic transport behavior of PANI nanoneedles synthesized by Dallas and coworkers was analyzed over a broad temperature range (10K–300K), with an interest focused on their metal–insulator transition that would be expected for a 1D semiconductor. In the same work, the authors studied the magnetic behavior of the materials. Even if the magnetization was expected to be weak, it would be of interest to correlate the magnetic and the electronic transport properties. In order to do that, magnetization versus temperature (M vs T, with a different external field applied) and magnetization versus magnetic field hysteresis curves was measured (M vs H).

As the physical properties of these materials depend on the electron delocalization and the potential presence of unpaired spins, the magnetic properties were correlated with electronic transport. At high temperature (where the resistance behavior was metallic), measurements indicated diamagnetic behavior similar to that for conventional metals. In contrast while lowering the temperature, the polymer started gradually to exhibit a paramagnetic behavior. Electron delocalization at high temperatures is responsible for such metallic and diamagnetic behavior and, as the temperature is lowered, the spins are localized and the order–disorder transition that takes place translates to paramagnetic and semiconductor–insulator behaviors, respectively. These transitions have been observed in other carbon nanomaterials, such as graphite filaments [66] and graphene monolayers. Graphene is also a well-known diamagnetic material and is defined as a "zero bandgap semiconductor" [67]. This behavior is also assigned to increased electron delocalization among the extended 2D-conjugated, sp2-hybridized carbon atoms that are forming a flat network. These are carbon microparticles or nanomaterials, but other organic molecules can demonstrate diamagnetic behavior, such as aromatics and fullerene "buckyballs." Fig. 4.8 shows a phase diagram constructed by Dallas and coworkers that is correlating the magnetic and electronic transport

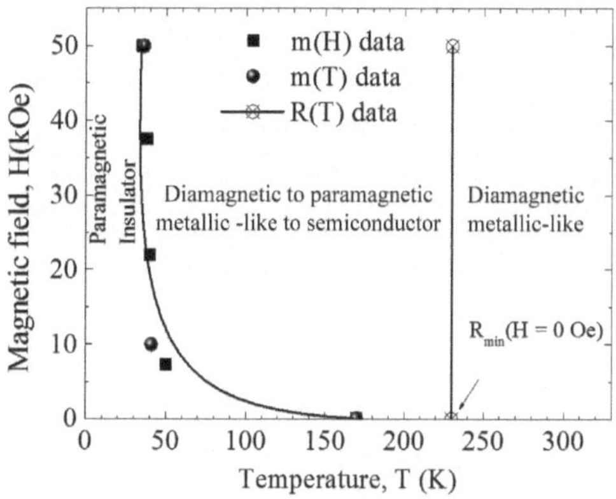

Figure 4.8 A phase diagram outlining the diamagnetic-to-paramagnetic and metal-to-insulator transitions for PANI nanoneedles synthesized through liquid–liquid interfaces. The acid used for the polymer initiation and doping was HCl and the immiscible organic phase was chloroform. *Adapted from A.J. Heeger, Resonance Raman Spectra and lattice dynamics calculations of pristine and doped polyconjugated polyfuran, Synth. Metals 55–57 (1993) 4371.*

properties of PANI nanoneedles (Fig. 4.8). This diagram was constructed based on electronic transport, M versus H, and M versus T measurements.

4.8 Other polymeric systems from interfaces

The phosphorus–nitrogen networks of cyclophosphazenes comprise an exciting class of materials that hold unprecedented opportunities for growing 3D structures through substitution of the chlorine groups. The chlorine groups are arranged in and out of the plane of the phosphonitrilic chloride trimer planar ring, hence 3D materials may be synthesized. The cyclophosphezenes have shown biocompatibility and they can form fluorescent particles for niche biomedical applications [68].

Examples of substituting ligands include aromatic diamines that lead to the formation of hollow tubes that can also reduce silver ions leading to silver nanocomposites with antibacterial activity [69]. These materials were synthesized through a refluxing toluene solution. On the other hand, thin cyclomatrix polyphosphazene films have been synthesized through interfacial polymerization of hexachlorocyclotriphosphazene ($P_3N_3Cl_6$) after reaction with an aromatic biphenol, while the pKa values of the biphenol used will influence the properties of the films. The morphology and the mechanical properties of the films depend on the density of the crosslink. Fig. 4.9 depicts the interfacial polymerization and cross-section SEM images of the thin films of the cyclophosphazene matrix. The phosphazene derivative from the interfacial polycondensation reaction is forming films on the surface of the polydimethylsiloxane substrate. The polydimethylsiloxane is immersed in the organic phase (in that case, dichloromethane) where the phosphonitrilic chloride is added along with the phase transfer catalyst which in that particular chase was the Aliquat 128. As a later step the films were transferred on polydimethylosiloxane stabs through dichloromethane solutions. Any excess of DCM solvent was removed and after the HCCP–biphenol film was completely dried, it could be removed from the PDMS surface.

As is already mentioned polyphosphazenes are used as flame retardants as one of their main applications. In recent works regarding PP composites, a high char yield has been observed in phosphazene derivatives and is attributed to extended crosslink [70]. Typical SEM images of the heated residues are shown in Fig. 4.10.

The liquid–vapor interface of ethanol at room temperature was studied by Sanyal et al through the scattering of X rays and was compared with the standard continuum capillary-wave theory. They calculated a root-mean-square fluctuation of the liquid surface at 6.9–0.2 A. The interface of liquid–vapor or liquid–liquid has long-ranged density correlations (i.e., macroscopic), while the correlations at a bulk liquid are short ranged as a consequence of dimensionality. Two-dimensional crystals of certain liquid crystal phases have displacement correlations that increase logarithmically as a function of distance, resulting in the loss of Bragg peaks. For liquid interfaces there is an analogous effect. It has been experimentally demonstrated that the air – liquid surface is extremely smooth with a root mean square roughness of about 3 Å. This mean square roughness corresponds to an air–water surface, in contrast to the bulk liquid, which is less ordered [71]. Two-dimensional polymers have been prepared on the macroscopic scale (in the range of cm^2 with a thickness of 1.2 nm and porosity of 6.8×1013 pores/cm^2), ~30% open area through the air–water interface where the monomer is restricted to two dimensions. β-Elaeostearin has been polymerized at the air–water interface in 1935 and later on the polymerization of calix[n]arenes at this interface. They demonstrated the large-scale synthesis of their material through polarized reflectance optical microscopy and scanning tunneling microscopy. Air–water interface and its nanometer resolution imaging. Trigonal star, amphiphilic monomers bearing three anthracene groups on a central

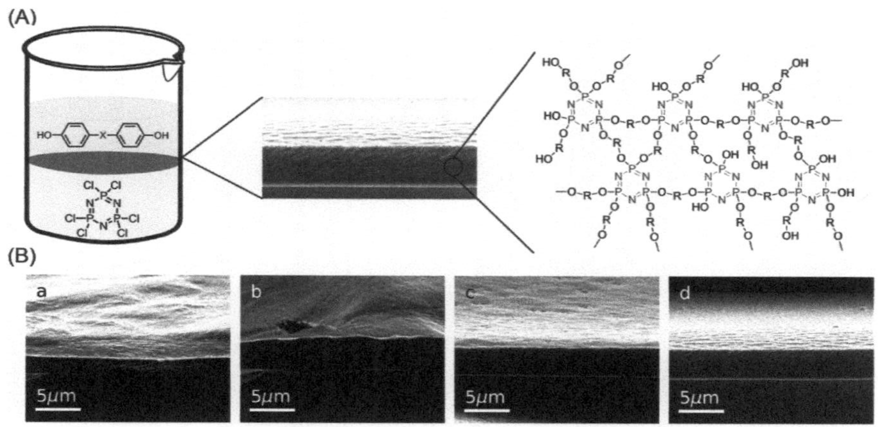

Figure 4.9 (A) A schematic representation of the interfacial polycondensation reaction between a phosphonitrilic chloride trimer core and ligands that are hydroxyl terminated. These —OH ligands substituted the chlorine atoms of the phosphazene core. (B) A cross-sectional scanning electron micrograph of the cyclophosphazene (HCCP)-biphenol films. (a) HCCP-BPH, (b) HCCP-DHPE, (c) HCCP-DHBP, and (d) HCCP-BPS. *Reproduced with permission by E. Maaskant, H. Gojzewski, M.A. Hempenius, G.J. Vancso, N.E. Benes. Thin cyclomatrix polyphosphazene films: interfacial polymerization of hexachlorocyclotriphosphazene with aromatic biphenols. Polym. Chem. 2018, 9, 3169–3180.*

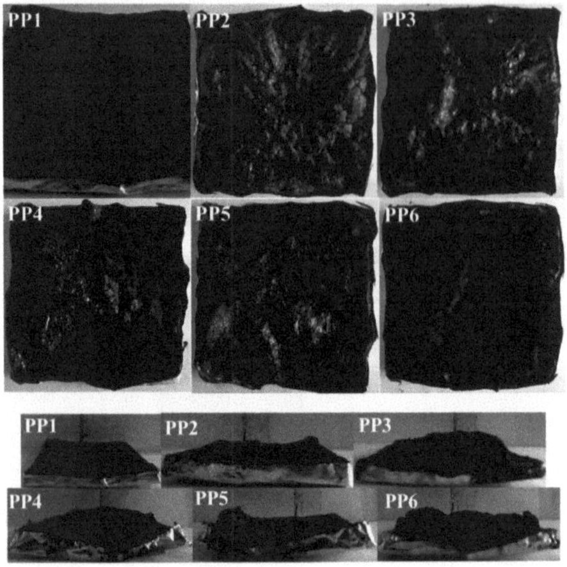

Figure 4.10 In these microscopy images are depicted charcoal residues from various polyphosphazene derivatives used as flame retardants. *Reproduced with permission from P. Wen, Q. Tai, Y. Hu, R.K.K. Yuen Cyclotriphosphazene-based intumescent flame retardant against the combustible polypropylene, Ind. Eng. Chem. Res. 55 (2016) 8018 – 8024.*

triptycene core are confined at the air–water interface. Compression followed by photopolymerization on the interface provides the 2D polymer [72]. An aqueous solution (10 mL) of $Cu(AcO)_2$ (2 mM) and pyridine (0.05 M) was added gently to the water phase. Subsequently the system was left for 3 days in the dark. The upper layer was then replaced with pure water, while the organic phase was replaced with pure CH_2Cl_2. After the removal of both aqueous and organic phases, the resultant film was suspended in ethanol. The suspension was cast dropwise on various substrates for measurement. An anthracene-based monolayer synthesized at the air–water interface having the strength to span 20×20 μm holes has been recently synthesized while strides toward a 2DP have been fabricated at the air–mercury interface.

Another novel and very interesting carbon allotrope is the graphdiyne (GDY). This one-atom-thick carbon allotrope exhibits layers of sp- and sp^2-hybridized carbon and can form nanowires, nanotube arrays, nanosheets, and ordered stripe arrays. It has been applied in catalysis, solar cells, energy storage, and optoelectronic devices [73]. The interfacial synthesis of 2D

N-graphdiyne films at the gas–liquid and liquid–liquid interfaces have also been reported. For that purpose triazine- or pyrazine-based monomers that had bearing ethynyl groups were polymerized through the Glaser coupling reactions at liquid–liquid interfaces and yielded multilayered or single-layered 2D *N*-graphdiyne materials with thin films with a minimum thickness of 4 nm [74]. An interesting and potentially important observation was that the wettability changed to more hydrophilic as the nitrogen content increased and the graphdiyne was further employed as a metal-free photocatalyst for NADH regeneration. It was found that the nitrogen content also affected the efficiency of the biocatalyst. For the synthesis of the materials, the upper, aqueous, layer contained copper(II) acetate and pyridine, which served as a catalyst for acetylenic homocoupling (which is the Eglinton coupling), and the lower dichloromethane layer contained the hexaethynylbenzene (now denoted as HEB) monomer. When dichloromethane was used as a solvent for the hexaethynylbenzene and subsequently was spread on the aqueous solution with the copper catalyst a multilayer graphdiyne was synthesized (24 nm; domain size, >25 μm). But through a gas–liquid system a single-crystalline graphdiyne with regular hexagonal domains and with less oxygen content was synthesized with a lateral size of 5 μm [75].

In these amine-containing polymers, similar to PANI, hydrogen bonds play a major role in their morphology, their intermolecular interactions and hence their physical properties. In Triazine-based sequence-defined polymers, recently the hydrogen bonding has recently showed that it can be exploited in a biomimetic fashion. By using molecular dynamics simulations, the triazine assemblies showed linear nanorods, that is, foldamers, by utilizing hydrogen bonds and $\pi-\pi$ interactions. The folding of hexamers into nanorods was much easier demonstrated when pentanediamine and *ortho*-xylenediamine bridges were placed in the center of the polymers. The foldamers can also undergo a *cis–trans* isomerization. Thus the experiments are providing evidence for a multifunctional new type of foldamer [76].

Polymer capsules are a class of polymeric materials with applications in pharmaceuticals, catalysis, and diagnostics. They are composed of a macromolecular shell and a hollow core with various external and internal diameters. For their synthesis, the processes of bottom-up self-assembly, layer-by-layer assembly, single-step polymer adsorption, and surface/interfacial polymerization are developed. Cui et al. [77] applied an interfacial polymerization for the synthesis of monodispersed polymer capsules with a size between 400 nm and 2.4 mm and a shell thickness from 10 to

140 nm. They further incorporated various molecules (thiocoraline), and NPs such as CdSe/CdS quantum dots (QDs), and magnetite Fe_3O_4 inside the polymer capsule. After the completion of the interfacial polymerization, the functionalized particles or the organic molecules will remain encapsulated inside the capsules.

In other works a controllable up-conversion (UC) system composed of uniform capsules has been fabricated through interfacial polymerization that was combined with a microfluidic approach using chromophore-dispersed monomer solutions. A photoinitiator and a surfactant were added in the aqueous liquid phase. At a subsequent step the capsule was illuminated with UV light and the free radicals that were produced in the aqueous phase were later diffused into the interface where the polymerization took place and the capsules finally had a fluidic core and an elastomeric layer. The shell thickness was estimated through electron microscopy on the order of ~10 nm and can be used for the encapsulation of drugs, dyes, and enzymes. The interfacial tension is bringing some limitations and so the particles that are synthesized through emulsion polymerization are in most cases spherical, with this picture changing during the interfacial polymerization where anisotropic Janus particles may be formed. In the work by Wang et al., anisotropic Janus particles have been extensively used as optical probes, solid surfactants, and self-propulsion motors and were fabricated via an oil-in-water emulsion, which contained hydrophobic monomers, hydrophilic monomers, and oil-soluble species [78]. These free-radical polymerizations are taking place exclusively to the interfaces due to the alternating nature of the copolymerization and the absence of homopolymerization between the two monomers, and no polymerization would be observed if the hydrophobic or the hydrophilic parts are absent. The emulsions were prepared from the following solutions: 10 wt.% oil phase composed of hexadecane, 25 wt.% dibutyl maleate (hydrophobic monomer), and 5% paraffin wax (hydrophobe), while the aqueous phase contained PEG-1000 divinyl ether (hydrophilic monomer), and 1 wt.% of dodecyltrimethylammonium bromide (DTAB) as surfactant. The system was polymerized at 60°C via injection of an interfacially active initiator, 2,2′-azobis(*N*-octyl-2-methylpropionamidine)dihydrochloride (3) (t1/2) 10 h at 59.5°C, dissolved in water and the particles were characterized with dynamic light scattering and electron microscopy (Fig. 4.11).

Wamser et al. [80] fabricated Janus porphyrin–polyamide composite films via interfacial polymerization. They employed tetrakis(4-aminophenyl)porphyrin (denoted as TAPP) or diethylenetriamine (DETA) with

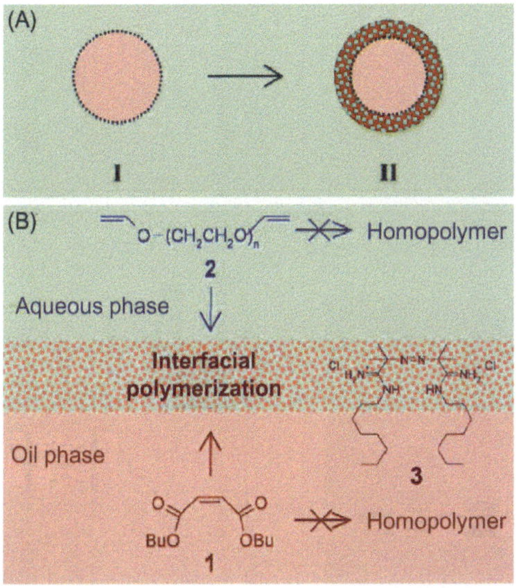

Figure 4.11 (A) Interfacial free-radical polymerization of oil drops (denoted as "I") which are forming liquid-core polymer capsules (denoted as "II"). The polymerization proceeds via (B) copolymerization of dibutyl maleate (denoted as "1") and the hydrophilic PEG divinyl ether (denoted as "2"). It is initiated through the surface-active initiator that is denoted as "3" at the oil–water interface. *Reproduced with permission from C. Scott, D. Wu, C.C. Ho, C.C. Co, Liquid-core capsules via interfacial polymerization: a free-radical analogy of the nylon rope trick, J. Am. Chem. Soc. 127 (2005) 4160–4161 [79].*

tetrakis[4-(chlorocarbonyl)phenyl]-porphyrin (denoted as TCCPP) separately dissolved in two immiscible liquids. The mechanism involves dechlorohydrogenation between the two monomers. Contact angle measurements demonstrated that the nitrogen side (TAPP/TCCPP film) was more hydrophobic than the carbon side (Fig. 4.12).

Another interesting class of materials that can be synthesized through interfacial polymerizations are the covalent organic frameworks (denoted as COFs). These crystalline polymers are bonded through covalent and nonionic bonds in 2D or 3D having a porosity as evidenced from BET. Matsumoto et al. reported an interfacial polymerization of an amine and an aldehyde monomer using a Lewis acid catalyst, the $Sc(OTf)_3$ which is a highly active catalyst for imine-linked COF formation [81]. By using immiscible solvents the authors segregated the catalyst from the monomers and thus obtained large-area COF films with an area of several cm^2 and thickness in the range of 100 mm to 2.5 nm.

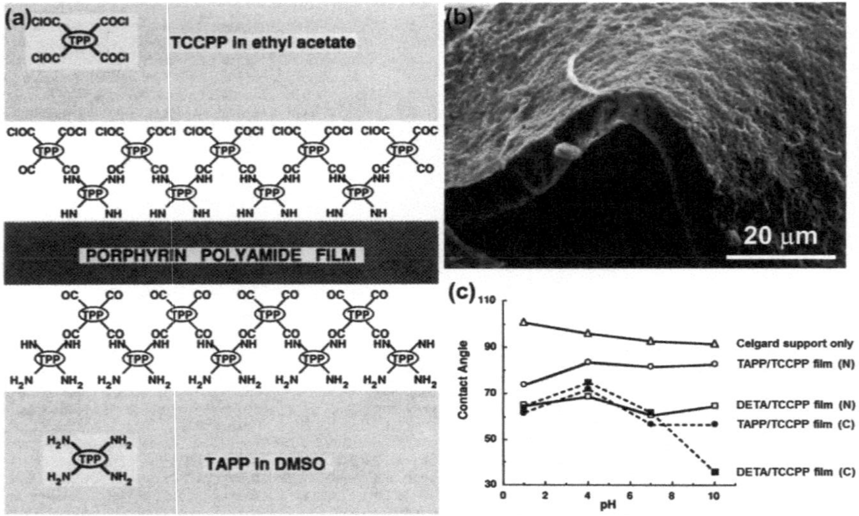

Figure 4.12 (A) A schematic describing the polymerization system with the TCCPP dissolved in ethyl acetate, the TAPP in DMSO and the porphyrin polyamide film at the interface. (B) An SEM image demonstrating the flexibility and ability for deformation of the film. (C) The nitrogen side of the TAPP/TCCPP film is proven to be more hydrophobic than the carbon side. *Reproduced with permission C.C. Wamser, R.R. Bard, V. Senthilathipan, V.C. Anderson, J.A. Yates, H.K. Lonsdale, et al., Synthesis and photoactivity of chemically asymmetric polymeric porphyrin films made by interfacial polymerization, J. Am. Chem. Soc. 111 (1989) 8485–8491. Copyright 1989, American Chemical Society.*

These films were subsequently transferred onto polyethersulfone substrates and they were tested toward Rhodamine WT. Reliable monomer design principles provide control of the COF's topology, pore size, and pore composition. COFs have found applications in gas adsorption, energy conversion, proton conductive films, catalysis, and storage [82]. The method has been sufficiently developed to yield layered materials with thickness even to a few nanometer or even a monolayer [83]. In general the monomers will polymerize into amorphous polymers that will grow into longer networks through imine exchange.

The noncatalytic liquid–liquid interfacial polymerizations required that the two monomers have orthogonal solubility. Furthermore the polymerizations under Langmuir–Blodgett conditions required amphiphilic monomers. In that case, the thickness of the film is dictated by the initial monomer concentration and ranges from 10 mm to 2.5 nm. The polymer films were then transferred to substrates. Other covalent organic frameworks films were transferred into polyethersulfone (denoted as PES)-supported nanofiltration (denoted as NF) membranes. The COFs on the NF membranes demonstrated the removal of Rhodamine pollutants [84]. The metal triflates, with a major example to be the Sc(OTf)3, have been repeatedly used for the synthesis of COFs (e.g., TAPB-PDA) that are derived from 1,3,5-tris(4-aminophenyl) benzene (TAPB) and terephthalaldehyde (PDA). In many experiments, acetic acid is used as a catalyst at high temperatures and long reaction times (e.g., 70 °C for72 h), but in contrast the $Sc(OTf)_3$-catalyzed polymerization can be accomplished in only 10–30 min at room temperature with low catalyst loading (<0.02 equiv). The imine-linked 2D materials are microcrystalline powders with a Brunauer–Emmett–Teller surface area of 2000 m^2/g. Adding water to the COF condensation reaction mixture results in the enhanced crystallinity of the powder when acetic acid was used as a catalyst. A layered 1,4-dioxane/mesitylene solution of TAPB and PDA monomers at various concentrations over an aqueous solution of Sc(OTf)3 formed films at the interface after 30 min with a thickness of 100 mm. With a 64-fold lower initial monomer concentration the polymerization progressed significantly slower and led to the formation of an interfacial film after 7 days that was transferred onto a silicon wafer. Profilometry studies revealed a thickness of 20 nm similar to previous works [85].

Other methods involve the addition of boronic acid where the monomer is prepolymerized through thermal treatment into nanocrystalline precursor COFs. A thermal treatment under reversible reaction conditions

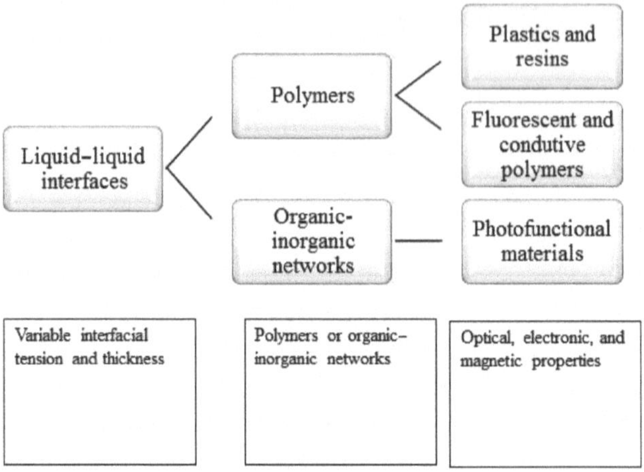

Figure 4.13 The different classes of polymeric materials that have been synthesized through liquid–liquid interfaces and their respective properties. Some typical polymer systems that have been derived from various liquid–liquid interfaces are presented.

facilitates the on-surface ripening and also shows a striking increase in the domain size [86].

4.9 Conclusion

The interfacial polymerization is an invaluable tool for polymer science and technology. It initiates, aids, and mediates the evolution of thin films, nanomaterials, and composites with light-emitting, electronic transport, antibacterial, and enhanced mechanical properties. Most studies have involved the synthesis of PANI fibers, but recent developments are moving toward organic electronic nanomaterials to 2D organic–inorganic nanosheets with interesting photophysical properties. The unique characteristics of the interfacial area may be applied for the "tailored" synthesis and design of functional polymers; hybrid networks with a thickness that can be tuned; composite materials with carbon or noble metal NP; highly conductive or fluorescent polymers with aromatic amines as the backbone. These polymeric materials, organic or hybrid organic–inorganic frameworks, and their respective properties are summarized in Fig. 4.10. This is a summary of the numerous different options available till date, and for future applications, the bespoke design of new monomers and building blocks will provide exciting new opportunities for functional materials (Fig. 4.13).

Ghosh and coworkers summarized, in a comprehensive and well-written review, the prospects of conductive polymer nanostructures in energy applications [87]. They highlighted the interfacial polymerization among other methods, such as soft or hard template assembly routes, the latter using anodized alumina and PC, radiolysis, and hydrogel formation. As an expansion of the discussion of the properties and applications of immiscible liquids, the role of the liquid–liquid interface in materials for energy applications is outlined in Chapter 6, Interfaces in Industrial Applications.

Despite their relatively recent discovery and development, the 2D materials have already found space for numerous practical applications due to the unusual delocalization of the electron cloud at the 2D space. Besides the well-known chemical vapor deposition methods, the 2D interfacial interface is a powerful tool for the development of functional materials. Specifically the light-emitting properties of various 2D materials are highly anticipated. However, they are dependent on environmental factors, optical sensors for temperature, stress/strain, gases, and solvents. In conclusion the parameters that influence the formation and properties of these materials deserve further investigations so that novel materials will be developed and their properties will be optimized [88].

References

[1] S.S. Dhumal, A.K. Suresh, Understanding interfacial polycondensation: experiments on polyurea system and comparison with theory, Polymer 51 (2010) 1176–1190.

[2] E.L. Wittbecker, P.W. Morgan, Interfacial polycondensation I, J. Polym. Sci., Part. A: Polym. Chem. Ed. 34 (1996) 521.

[3] H.D. Tran, J.M. D'Arcy, Y. Wang, P.J. Beltramo, V.A. Strong, R.B. Kaner, The oxidation of aniline to produce "polyaniline": a process yielding many different nanoscale structures, J. Mater. Chem. 21 (2011) 3534.

[4] a) S. Edward, J.-W. Choi, Conducting polyaniline nanowire and its applications in chemiresistive sensing, Nanomaterials 3 (2013) 498.
b) J. Huang, R.B. Kaner, Angew. Chem. Int. Ed. 43 (2004) 5817.
c) J. Huang, R.B. Kaner, Polyaniline nanowires on Si surfaces fabricated with DNA templates, J. Am. Chem. Soc. 126 (2004) 851.
d) D. Li, J. Huang, R.B. Kaner, Polyaniline nanofibers: a unique polymer nanostructure for versatile applications, Acc. Chem. Res. 42 (2009) 133.

[5] J. Chen, D. Chao, X. Lu, W. Zhang, Novel interfacial polymerization for radially oriented polyaniline nanofibers, Mater. Lett. 61 (2007) 1419.

[6] M. Moniruzzaman, J. Chattopadhyay, W.E. Billups, K.I. Winey, Tuning the mechanical properties of SWNT/Nylon 6,10 composites with flexible spacers at the interface, Nano. Lett. 7 (2007) 1178.

[7] R. Haggenmueller, F. Du, J.E. Fischer, K.I. Winey, Interfacial in situ polymerization of single wall carbon nanotube/nylon 6,6 nanocomposites, Polymer 47 (2006) 2381–2388.
[8] H. Ichiura, T. Seike, Y. Ohtani, Interfacial polymerization of functional paper: morphology of the nylon film prepared on paper surfaces, Ind. Eng. Chem. Res. 52 (2013) 9137–9144.
[9] N. Ning, S. Fu, W. Zhang, F. Chen, K. Wang, H. Deng, et al., Realizing the enhancement of interfacial interaction in semicrystalline polymer/filler composites via interfacial crystallization, Prog. Polym. Sci. 37 (2012) 1425–1455.
[10] A. Mulchandani, N.V. Myung, Conducting polymer nanowires-based label-free biosensors, Curr. Opin. Biotechnol. 22 (2011) 502.
[11] J. Huang, R.B. Kaner, Nanofiber formation in the chemical polymerization of aniline: a mechanistic study, Angew. Chem. Int. Ed. 43 (2004) 5817.
[12] S. Edward, J.-W. Choi, Conducting polyaniline nanowire and its applications in chemiresistive sensing, Nanomaterials. 3 (2013) 498.
[13] a) W. Li, Q. Zhang, D. Chen, L. Li, Study on nanofibers of polyaniline via interfacial polymerization, J. Macromol. Sci. Part A 43 (2006) 1815.
b) H. Gao, T. Jiang, B. Han, Y. Wang, J. Du, Z. Liu, et al., Polymer 45 (2004) 3017.
c) Y. He, One-dimensional polyaniline nanostructures synthesized by interfacial polymerization in a solids-stabilized emulsion, Appl. Surf. Sci. 252 (2006) 2115.
[14] J. Zhang, D. Mandler, P.R. Unwin, Interfacial polymerization of anilinium at Langmuir monolayers, Chem. Commun. 450 (2004).
[15] J. Li, Q. Jia, J. Zhu, M. Zheng, Synthesis of polyaniline particles via inverse and direct miniemulsion, Polym. Int. 57 (2008) 337.
[16] J.M. Neugebauer, Detergents: an overview, Meth. Enzymol. 182 (1990) 239.
[17] S. Liu, K. Zhu, Y. Zhang, J. Xu, Polymer 47 (2006) 7680.
[18] S. Ding, H. Mao, W. Zhang, Fabrication of DBSA-doped polyaniline nanorods by interfacial polymerization, J. Appl. Polym. Sci. 109 (2008) 2842.
[19] T.V. Vernitskaya, O.N. Efimov, Polypyrrole: a conducting polymer; its synthesis, properties and applications, Russ. Chem. Rev. 66 (1997) 443.
[20] P. Dallas, V. Georgakilas, Interfacial polymerization of conductive polymers: generation of polymeric nanostructures in a 2-D space, Adv. Colloid Interf. Sci. 224 (2015) 46.
[21] A.J. Pomposo, E. Ochoteco, C. Pozo, P.-M. Carrasco, H.-J. Grande, F.J. Rodriguez, Conductivity enhancement in raw polypyrrole and polypyrrole nanoparticle dispersions, Polym. Adv. Technol. 17 (2006) 26.
[22] S. Shang, W. Zeng, X. Tao, Fabrication of conducting polypyrrole/β-cyclodextrin nano- and micro-spheres using molecular templates, RSC Adv. 2 (2012) 4675.
[23] K.P. Lee, A.I. Gopalan, S.H. Lee, M.S. Kim, Polyaniline and cyclodextrin based chiral nanobundles—functional materials having size and enantioselectivity, Nanotechnology 17 (2006) 375.
[24] a) Y. Tan, K. Ghandi, Kinetics and mechanism of pyrrole chemical polymerization, Synth. Metals. 175 (2013) 183.
b) P. Dallas, D. Niarchos, D. Vrbanic, N. Boukos, S. Pejovnik, D. Petridis, Polypyrrole/MWNT nanocomposites synthesized through interfacial polymerization, Polymer 48 (2007) 2007.
c) J. Stejskal, Conducting polymer–silver composites, Chem. Papers 67 (2013) 814.
d) L.M. Huang, W.H. Liao, H.C. Ling, T.C. Wen, Simultaneous synthesis of polyaniline nanofibers and metal (Ag and Pt) nanoparticles, Mater. Chem. Phys. 116 (2009) 474.

e) M.D. Bedre, S. Basavaraja, B.D. Salwe, V. Shivakumar, L. Arunkumar, A. Venkataraman, Preparation and characterization of Pani and Pani-Ag nanocomposites via interfacial polymerization, Polym. Compos. 30 (2009) 1668.
[25] P. Mukherjee, A.K. Nandi, Growth of different shape au nanoparticles through an interfacial redox process using a conducting polymer, Langmuir 26 (2010) 2785.
[26] F. Yang, Y. Chu, S. Ma, Y. Zhang, J. Liu, Preparation of uniform silica/polypyrrole core/shell microspheres and polypyrrole hollow microspheres by the template of modified silica particles using different modified agents, J. Colloid Interf. Sci. 301 (2006) 470.
[27] A.L. Kukla, Y.M. Shirshov, S.A. Piletsky, Ammonia sensors based on sensitive polyaniline film, Sens. Actuat. B 37 (1996) 135.
[28] a) Y. Tan, K. Ghandi, Kinetics and mechanism of pyrrole chemical polymerization, Synth. Metals 175 (2013) 183.
b) J. Stejskal, Chem. Papers 67 (2013) 814.
c) L.M. Huang, W.H. Liao, H.C. Ling, T.C. Wen, Mater. Chem. Phys. 116 (2009) 474.
d) M.D. Bedre, S. Basavaraja, B.D. Salwe, V. Shivakumar, L. Arunkumar, A. Venkataraman, Polym. Compos. 30 (2009) 1668.
[29] X.B. Yan, Z.J. Han, Y. Yang, B.K. Tay, NO_2 gas sensing with polyaniline nanofibers synthesized by a facile aqueous/organic interfacial polymerization, Sens. Actuat. B 123 (2007) 107.
[30] a) U. Tamer, C. Kanbes, N. Erta, Branched fibers of conducting polypyrrole: synthesis and characterization, Intern. J. Polym Anal. Charact. 14 (2009) 259.
b) U.M. Fatema, M.M. Rahman, M.R. Islam, M.Y.A. Mollah, M.A.B.H. Susan, Silver/poly(vinyl alcohol) nanocomposite film prepared using water in oil microemulsion for antibacterial applications, Colloid Interf. Sci. 514 (2018) 648–655.
[31] Y. Yan, J. Cui, S. Zhao, J. Zhang, J. Liu, J. Cheng, Interface molecular engineering of single-walled carbon nanotube/epoxy composites, J. Mater. Chem. 22 (5) (2012) 1928.
[32] T. Hu, L. He, L. Duan, Y. Qiu, Solid-state light-emitting electrochemical cells based on ionic iridium(iii) complexes, J. Mater. Chem. 16 (2006) 4206.
[33] J.-H. Shi, B.-X. Yang, K.P. Pramoda, S.H. Goh, Enhancement of the mechanical performance of poly(vinyl chloride) using poly(n -butyl methacrylate)-grafted multi-walled carbon nanotubes, Nanotechnology 18 (37) (2007) 375704.
[34] Z. Spitalsky, D. Tasis, K. Papagelis, C. Galiotis, Carbon nanotube–polymer composites: chemistry, processing, mechanical and electrical properties, Prog. Polym. Sci. 35 (2010) 357.
[35] U. Tamer, C. Kanbes, N. Erta, Branched fibers of conducting polypyrrole: synthesis and characterization, Int. J. Polym. Anal. Charact. 14 (2009) 259.
[36] a) A.J. Pomposo, E. Ochoteco, C. Pozo, P.-M. Carrasco, H.-J. Grande, F.J. Rodriguez, Conductivity enhancement in raw polypyrrole and polypyrrole nanoparticle dispersions, Polym. Adv. Technol. 17 (2006) 26.
b) V. Georgakilas, P. Dallas, D. Niarchos, N. Boukos, C. Trapalis, Polypyrrole/MWNT nanocomposites synthesized through interfacial polymerization, Synth. Metals 159 (2009) 632.
[37] A. Deepshikha, T. Basu, A review on synthesis and characterization of nanostructured conducting polymers (NSCP) and application in biosensors, Anal. Lett. 44 (2011) 1126.
[38] M. Joulazadeh, A.H. Navarchian, M. Niroomand, A comparative study on humidity sensing performances of polyaniline and polypyrrole nanostructures, Adv. Polym. Technol. 33 (2014) 21461.

[39] J. Huang, S. Virji, R.B. Kaner, B.H. Weiller, Nanostructured polyaniline sensors, Chem. a Eur. J. 10 (2004) 1314.

[40] V.I. Krinichnyi, S.V. Tokarev, H.K. Roth, M. Schrodner, B. Wessling, EPR study of charge transfer in polyaniline highly doped by p-toluenesulfonic acid, Synth. Met. 156 (2006) 1368.

[41] S. Virji, J. Huang, R.B. Kaner, B.H. Weiller, Polyaniline nanofiber gas sensors: examination of response mechanisms, Nano. Lett. 4 (2004) 491.

[42] a) A.L. Kukla, Y.M. Shirshov, S.A. Piletsky, Sensors Actuat. B 37 (1996) 135.
b) L. Wang, H. Huang, S. Xiao, D. Cai, Y. Liu, B. Liu, et al., ACS Appl. Mater. Interf. 6 (2014) 14131.
c) S. Wu, F. Zeng, F. Li, Y. Zhu, Eur. Polym. J. 36 (2000) 679.

[43] S. Virji, J.D. Fowler, C.O. Baker, J.X. Huang, R.B. Kaner, B.H. Weiller, Polyaniline manofiber composites with metal salts: chemical sensors for hydrogen sulfide, Small 1 (2005) 624.

[44] H.J. Kim, S. Im, J.C. Kim, W.G. Hong, K. Shin, H.Y. Jeong, et al., Phytic acid doped polyaniline nanofibers for enhanced aqueous copper(II) adsorption capability, ACS Sustain. Chem. Eng. 5 (2017) 6654–6664.

[45] L. Wang, H. Huang, S. Xiao, D. Cai, Y. Liu, B. Liu, et al., ACS Appl. Mater. Interf. 6 (2014) 14131.

[46] F.G. Souza, G.E. Oliveira, T. Anzai, P. Richa, T. Cosme, M. Nele, et al., Macromol. Mater. Eng. 294 (2009) 739.

[47] a) Y. Xia, J.M. Wiesinger, A.G. MacDiarmid, Chem. Mater. 7 (1995) 443.
b) J.M. Davey, C.O. Too, S.F. Ralph, L.K. Maguire, G.C. Wallace, Macromolecules 33 (2000) 7044.
c) W.S. Huang, B.D. Humphrey, A.G. MacDiarmid, J. Chem. Soc. Faraday Transact. 82 (1986) 2385.

[48] a) S.C. Hernandez, D. Chaudhuri, W. Chen, N.V. Myung, A. Mulchandani, Electroanalysis 19–20 (2007) 2125.
b) N. Nuraje, K. Su, N.L. Yang, H. Matsui, ACS Nano 2 (2008) 502–2506.
K. Su, N. Nuraje, L.Z. Zhang, I.W. Chu, R.M. Peetz, H. Matsui, et al., Adv. Mater. 19 (2007) 669–672.
c) D.J. Li, Y. Li, Y.Y. Feng, W.P. Hu, W. Feng, J. Mater. Chem. A 3 (2015) 2135–2143.
d) Y.S. Choi, S.H. Joo, S.A. Lee, D.J. You, H. Kim, C. Pak, et al., Macromolecules 39 (2006) 3275–3282.

[49] a) Y.B. Moon, Y. Cao, P. Smith, A.J. Heeger, Polym. Commun. 30 (1989) 196.
b) C.H. Yang, Y.K. Chih, H.E. Cheng, C.H. Chen, Polymer 46 (2005) 10688.

[50] E. Detsri, S.T. Dubas, J. Appl. Polym. Sci. 128 (2013) 558.

[51] a) N. Nuraje, K. Su, N.I. Yang, H. Matsui, ACS Nano 2 (2008) 502.
b) K. Su, N. Nuraje, L. Zhang, I.W. Chu, R.M. Peetz, H. Matsui, et al., Adv. Mater. 19 (2007) 669.

[52] a) M.C. Park, Q. Sun, Y. Deng, Macromol. Rapid Commun. 28 (2007) 1237.
b) J.P. Pouget, M.A. Jozefowicz, A.J. Epstein, X. Tang, A.G. MacDiarmid, Macromolecules 24 (1991) 779.

[53] A. Manim, K. Athinarayanasamy, P. Kamaraj, J. Mater. Sci. Lett. 14 (1995) 1594.

[54] W. Zheng, M. Angelopoulos, A.J. Epstein, A.G. MacDiarmid, Macromolecules 30 (1997) 7634.

[55] A.G. MacDiarmid, A.J. Epstein, Synth. Met. 69 (1995) 85.

[56] N.T.K. Thanh, N. Maclean, S. Mahiddine, Chem. Rev. 114 (2014) 7610.

[57] J. Tsukamoto, Adv. Phys. 41 (1992) 509.

[58] A.B. Kaiser, Rep. Progress Phys. 64 (2001) 1.
b) Z. Wei, C.F. Faul, Macromol. Rapid Commun. 29 (2008) 280.

[59] C. Su, G. Wang, F. Huang, X. Li, J. Mater. Sci. 43 (2008) 197.

[60] G. Strobl, 287 The Physics of Polymers: Concepts for Understanding Their Structures and Behavior, Springer-Verlag Berlin, Heidelberg, 2007.

[61] S. Stafstrom, J.L. Bredas, A.J. Epstein, H.S. Woo, D.B. Tanner, W.S. Huang, et al., Polaron lattice in highly conducting polyaniline – theoretical and optical studies, Phys. Rev. Lett. 59 (1987) 1464.

c) J. Stejskal, P. Kratochvil, Polyaniline dispersions 2. UV—Vis absorption spectra, Synth. Metals 61 (1993) 225.

[62] a) A.J. Heeger, Resonance Raman Spectra and lattice dynamics calculations of pristine and doped polyconjugated polyfuran, Synth. Metals 55–57 (1993) 4371.

b) H. Snalwa, Advanced Functional Molecules and Polymers: Electronic and Photonic Properties: Electronic and Photonic Properties, vol. 3, Taylor and Francis, Abingdon, 2011, p. 130.

[63] a) P. Dallas, D. Stamopoulos, N. Boukos, V. Tzitzios, D. Niarchos, D. Petridis, Characterization, magnetic and transport properties of polyaniline synthesized through interfacial polymerization, Polymer 48 (2007) 3162.

b) J. Unsworth, B.A. Lunn, P.C. Innis, Z. Jin, A. Kaynak, N.G. Booth, Technical review: conducting polymer electronics, J. Intell. Mater. Syst. Struct. 3 (1992) 380.

c) M. Angelopoulos, Conducting polymers in microelectronics, IBM J. Res. Dev. 45 (2001) 57.

d) N. Gospodinova, L. Terlemezyan, Prog. Polym. Sci. 23 (1998) 1443.

e) S. Bhadra, S. Chattopadhyay, N.K. Singha, D. Khastgir, Change in fiber properties due to the heat treatment of nylon 6 tire cords, J. Appl. Polym. Sci. 108 (2008) 57.

[64] a) N.F. Mott, D.A. David, Electronic Processes in Non-Crystalline Materials, Oxford University Press, Oxford, 1979.

b) M. Reghu, Y. Cao, D. Moses, A.J. Heeger, Phys. Rev. B 47 (1993) 1758.

[65] a) K. Lee, K.S. Cho, S.H. Sung, A.J. Heeger, C.W. Lee, S.H. Lee, Nature 41 (2006) 65.

b) S.H. Lee, D.H. Lee, K. Lee, C.W. Lee, Adv. Funct. Mater. 15 (2005) 1495.

c) B. Ruzicka, L. Degiorgi, Phys. Rev. B 61 (2000) R2468.

[66] a) J.C. Calderon-Moreno, A. Labarta, X. Batlle, D. Crespo, V.G. Pol, S.V. Pol, et al., Carbon 44 (2006) 2864.

b) M.P. Johansson, J. Juselius, D. Sundholm, Angew. Chem. Int. Ed. 44 (2005) 1843.

b) Y. Shibayama, H. Sato, T. Enoki, Disordered magnetism at the metal-insulator threshold in nano-graphite-based carbon materials, Phys. Rev. Lett. 84 (2000) 1744.

c) K. Wabayashi, M. Fujita, H. Ajiki, M. Sigrist, Electronic and magnetic properties of nanographite ribbons, Phys. Rev. B 59 (1999) 8271.

[67] a) H. Possel, H. Noack, W. Augustin, G. Keilhoff, G. Wolf, 2,7-Dihydrodichlorofluorescein diacetate as a fluorescent marker for peroxynitrite formation, FEBS Lett 416 (2) (1997) 175.

b) L.J. Sun, T.H. Liu, H. Li, L. Yang, L.J. Meng, Q.H. Lu, et al., Fluorescent and cross-linked organic-inorganic hybrid nanoshells for monitoring drug delivery, ACS Appl Mater Interf 7 (8) (2015) 4990.

c) L. Zhu, Y. Zhu, Y. Pan, Y. Huang, X. Huang, X. Tang, Fully crosslinked poly [cyclotriphosphazene-co-(4,4′-sulfonyldiphenol)] microspheres via precipitation polymerization and their superior thermal properties, Macromol. React. Eng. 1 (1) (2007) 45–52.

[68] A.L. Kuzemsky, Generalized kinetic and evolution and equations in the approach of the nonequilibrium statistical operator, Int. J. Mod. Phys. B 27 (2013) 1330007.

[69] a) P. Dallas, J. Tucek, D. Jancik, M. Kolar, A. Panacek, R. Zboril, Magnetically controllable silver nanocomposite with multifunctional phosphotriazine matrix and high antimicrobial activity, Adv. Funct. Mater. 20 (2010) 2347.
b) P. Dallas, R. Zboril, A.B. Bourlinos, D. Jancik, D. Niarchos, A. Panacek, et al., Cornet-like phosphotriazine/diamine polymers as reductant and matrix for the synthesis of silver nanocomposites with antimicrobial activity, Macromol. Mater. Eng. 295 (2010) 108.

[70] P. Wen, Q. Tai, Y. Hu, R.K.K. Yuen, Cyclotriphosphazene-based intumescent flame retardant against the combustible polypropylene, Ind. Eng. Chem. Res. 55 (2016) 8018–8024.

[71] M.K. Sanyal, S.K. Sinha, K.G. Huang, B.M. Ocko, X-ray scattering study of capillary-wave fluctuations at a liquid surface, Phys. Rev. Lett. 66 (1991) 628–631.
D.J. Murray, D.D. Patterson, P. Payamyar, R. Bhola, W. Somg, M. Lackinger, et al., Large area synthesis of a nanoporous two-dimensional polymer at the air/water interface, J. Am. Chem. Soc 137 (2015) 3450.

[72] X. Kan, Y. Ban, C. Wu, O. Pan, H. Liu, J. Song, et al., Interfacial synthesis of conjugated two-dimensional N-graphdiyne, ACS Appl. Mater. Interf. 10 (1) (2018) 53–58.

[73] H. Yu, Y. Xue, Y. Li, Graphdiyne and its assembly architectures: synthesis, functionalization, and applications, Adv. Mater. 31 (2019) 1803101.

[74] Q. Pan, H. Liu, Y. Zhao, S. Chen, B Xue, X. Kan, et al., Preparation of N-graphdiyne nanosheets at liquid/liquid interface for photocatalytic NADH regeneration, ACS Appl. Mater. Interf. 11 (2019) 2740–2744.

[75] R. Matsuoka, R. Sakamoto, K. Hoshiko, S. Sasaki, H. Masunaga, K. Nagashio, et al., Crystalline graphdiyne nanosheets produced at a gas/liquid or liquid/liquid interface, J. Am. Chem. Soc. 139 (2017) 3145–3152.

[76] a) A.K. Ghosh, B.H. Jeong, X. Huang, E.M.V. Hoek, Impacts of reaction and curing conditions on polyamide composite reverse osmosis membrane properties, J. Membr. Sci. 311 (2008) 34.
b) S. Ghosh, T. Maiyalagan, R.N. Basu, Nanostructured conducting polymers for energy applications: towards a sustainable platform, Nanoscale 8 (2016) 6921.

[77] J.W. Cui, Y.J. Wang, A. Postma, J.C. Hao, L. Hosta-Rigau, F. Caruso, Monodisperse polymer capsules: tailoring size, shell thickness, and hydrophobic cargo loading via emulsion templating, Adv. Funct. Mater. 20 (2010) 1625–1631.

[78] a) A. Walther, A.H.E. Mueller, Janus particles: synthesis, self-assembly, physical properties, and applications, Chem. Rev 113 (2013) 5194–5261.
b) F. Liang, K. Shen, X. Qu, C. Zhang, Q. Wang, J. Li, et al., Inorganic janus nanosheets, Angew. Chem., Int. Ed. 50 (2011) 2379–2382.
c) F. Mou, C. Chen, H. Ma, Y. Yin, Q. Wu, J. Guan, Self-propelled micromotors driven by the magnesium–water reaction and their hemolytic properties, Angew. Chem., Int. Ed. 52 (2013) 7208–7212.
d) J.-B. Fan, Y. Song, H. Li, J.-P. Jia, X. Guo, L. Jiang, Controllable drug release and effective intracellular accumulation highlighted by anisotropic biodegradable PLGE nanoparticles, J. Mater. Chem. B 2 (2014) 3911–3914.

[79] C. Scott, D. Wu, C.C. Ho, C.C. Co, Liquid-core capsules via interfacial polymerization: a free-radical analogy of the nylon rope trick, J. Am. Chem. Soc. 127 (2005) 4160–4161.

[80] C.C. Wamser, R.R. Bard, V. Senthilathipan, V.C. Anderson, J.A. Yates, H.K. Lonsdale, et al., Synthesis and photoactivity of chemically asymmetric polymeric porphyrin films made by interfacial polymerization, J. Am. Chem. Soc. 111 (1989) 8485–8491.

[81] M. Matsumoto, L. Valentino, G.M. Stiehl, D.C. Ralph, B.J. Marinas, W.R. Dichtel, Covalent organic frameworks: from lewis-acid-catalyzed interfacial polymerization of covalent organic framework films, Chem 4 (2018) 308–317.
[82] a) X. Feng, X.X. Ding, D.D. Jiang, Covalent organic frameworks, Chem. Soc. Rev. 41 (2012) 6010–6022.
b) S.Y. Ding, W. Wang, Covalent organic frameworks (COFs): from design to applications, Chem. Soc. Rev. 42 (2013) 548–568.
c) C.J. Doonan, D.J. Tranchemontagne, T.G. Glover, J.R. Hunt, O.M. Yaghi, Exceptional ammonia uptake by a covalent organic framework, Nat. Chem. 2 (2010) 235–238.
d) M.G. Rabbani, A.K. Sekizkardes, Z. Kahveci, T.E. Reich, R. Ding, H.M. El-Kaderi, A 2D mesoporous imine-linked covalent organic framework for high pressure gas storage applications, Chem. Eur. J. 19 (2013) 3324–3328.
C.R. DeBlase, K.E. Silberstein, T.T. Truong, H.D. Abruna, W.R. Dichtel, Beta Ketoenamine-linked covalent organic frameworks capable of pseudocapacitive energy storage, J. Am. Chem. Soc. 135 (2013) 16821–16824.
e) S.Y. Ding, J. Gao, Q. Wang, Y. Zhang, W.G. Song, C.Y. Su, et al., Construction of covalent organic framework for catalysis: Pd/COF-LZU1 in Suzuki-Miyaura coupling reaction, J. Am. Chem. Soc. 133 (2011) 19816.
f) J.W. Colson, A.R. Woll, A. Mukherjee, M.P. Levendorf, E.L. Spitler, V.B. Shields, et al., Oriented 2D covalent organic framework thin films on single-layer graphene, Science 332 (2011) 228–231.
g) X.H. Liu, C.Z. Guan, D. Wang, L.J. Wan, Graphene-like single-layered covalent organic frameworks: synthesis strategies and application prospects, Adv. Mater. 26 (2014) 6912–6920.
[83] a) J.L. Feldblyum, C.H. McCreery, S.C. Andrews, T. Kurosawa, E.J. Santos, V. Duong, et al., Few-layer, large-area, 2D covalent organic framework semiconductor thin films, Chem. Commun. 51 (2015) 13894–13897.
b) B.J. Smith, A.C. Overholts, N. Hwang, W.R. Dichtel, Insight into the crystallization of amorphous imine-linked polymer networks to 2D covalent organic frameworks, Chem. Commun. 52 (2016) 3690–3693.
[84] Z. Zheng, R. Grunker, X. Feng, Synthetic two-dimensional materials: a new paradigm of membranes for ultimate separation, Adv. Mater. 28 (2016) 6529–6545.
[85] R. Sakamoto, T. Yagi, K. Hoshiko, S. Kusaka, R. Matsuoka, H. Maeda, et al., Photofunctionality in porphyrin-hybridized Bis(dipyrrinato)zinc(II) complex micro- and nanosheets, Angew. Chem. Int. Ed. 56 (2017) 3526–3530.
[86] J.F. Dienstmaier, A.M. Gigler, A.J. Goetz, P. Knochel, T. Bein, A. Lyapin, et al., Synthesis of well-ordered COF monolayers: surface growth of nanocrystalline precursors versus direct on-surface polycondensation, ACS Nano 512 (2011) 9737–9745.
[87] a) Z. Zhang, F. Sun, Z. Zhu, J. Dai, K. Gao, Q. Wei, et al., Unconventional solution-phase epitaxial growth of organic-inorganic hybrid perovskite nanocrystals on metal sulfide nanosheets, Sci. Chin. Mater. 62 (2019) 43.
b) C. Tan, Z. Zeng, X. Huang, X. Rui, X.J. Wu, B. Li, et al., Liquid-phase epitaxial growth of two-dimensional semiconductor hetero-nanostructures, Angew. Chem. Int. Ed. 54 (2015) 1841–1845.
C. Tan, J. Chen, X.-J. Wu, H. Zhang, Epitaxial growth of hybrid nanostructures, Nat. Rev. Mater 3 (2018) 17089.
e) S. Ghosh, T. Maiyalagan, R.N. Basu, Nanostructured conducting polymers for energy applications: towards a sustainable platform, Nanoscale 8 (2016) 6921.
[88] S.-H. Ahn, J.W. Grate, Foldamer architectures of triazine-based sequence-defined polymers investigated with molecular dynamics simulations and enhanced sampling methods, J. Phys. Chem. B. 123 (2019) 9364.

CHAPTER 5

Light-emitting materials generated at the liquid–liquid interface

5.1 Introduction

Light-emitting organic compounds comprise a very important class of materials that are strong candidates for various applications such as sensors, nanodevices, bioimaging and optoelectronics photonics, sensors, and biomarkers in cell imaging [1]. The most typical organic fluorescent materials include small molecules such as fluorescein, rhodamine B, and anthracene. However, nanoparticles such as quantum dots and macromolecules with carefully selected functional groups will lead to the formation of novel fluorophores and even push–pull dyes with long wavelength emissions and long fluorescence lifetimes. Other examples of fluorescent materials include carbon dots, green fluorescent proteins, inorganic quantum dots, and lanthanide chelates [2]. Carbon dots are a particularly interesting, and relatively new class of materials, with excellent solubility in water and remarkable stability in extreme environments, such as underground water streams where the pressure, temperature, and salinity are very high. Carbon dots are also very easy to be synthesized, usually through a straightforward thermal treatment of an organic amine and an organic base, hence a scaled-up production can be achieved with low cost and combing biocompatibility and scalability. The ease of preparation, low cost, and remarkable stability make these carbogenic nanoparticles ideal candidates for application in oil tracing in underground environments. This stability inharsh environments and underground conditions is a property that few nanomaterials possess. The Giannelis group has pioneered the use of carbon dots as reservoir nanosized agents for oil tracing [3a]. While other nanoparticles are very stable in porous media; they lack stability when passing through positively charged calcite [3b]. Furthermore carbon dots do not possess the toxicity of the cadmium-based CdSe quantum dots. Their emission is usually located in the blue region, with emission at $\lambda = 470$ nm after excitation at $\lambda = 375$ nm, however after

Generation of Polymers and Nanomaterials at Liquid-Liquid Interfaces
DOI: https://doi.org/10.1016/B978-0-12-819491-1.00005-8

suitable doping, for example, fluorine doping, they can possess a red-shifted emission [3c].

On the other hand, conductive polymers, a class of materials that was thoroughly analyzed in t Chapter 4, Synthesis and Generation of Polymeric Materials From Interfaces, in general suffer from very low fluorescence quantum and poor optical properties something that hinders their potential applications as optoelectronics components and fluorescent sensors [4a]. In this chapter, the ability to change this behavior through substitution with perfluorinated pendant groups will be analyzed.

Depending on the desired application, a different optimum emission wavelength will be anticipated. Examples include telecommunications and biomedical applications where an infrared emission is desirable. To that end, rare earth metal cations possess an exciting infrared phosphorescence behavior and they can be incorporated in metal–organic frameworks (MOFs) or even carbon nanomaterials. Erbium is a characteristic example and has been incarcerated in carbon nanomaterials while still retaining its exciting optical properties. For example, the infrared emission of the erbium containing fullerenes is invaluable for their use as silicon fiber amplifiers due to the location of the λ_{max} at 1520 nm, which is located at the telecommunications window.

Fig. 5.1 presents a summary of the emission maxima of a series of materials. This graph includes the carbon dots with a blue fluorescence at 470 nm, quantum dots such as CdSe, fullerenes, carbon nanotubes, and erbium incarcerating endohedral metallofullerenes. The latter comprises

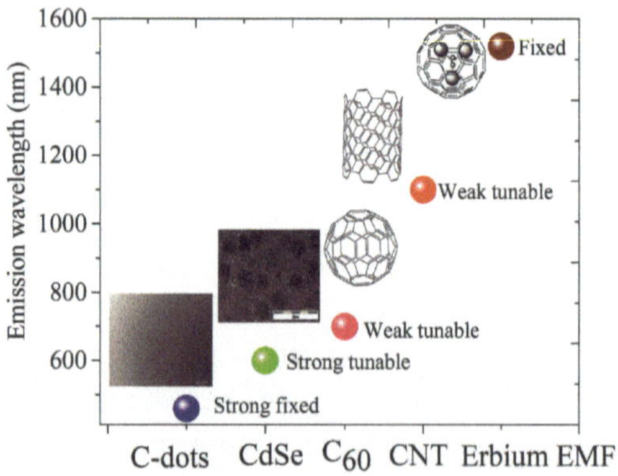

Figure 5.1 A summary of important light-emitting nanomaterials and their emission maximum, λ_{max}. The examples shown are carbon dots, quantum dots, fullerenes, carbon nanotubes, and erbium incarcerating endohedral metallofullerenes.

an exciting class of materials where the erbium ion emits light at the infrared region and is protected inside a carbon cage. This cage can be subsequently functionalized, something that opens new pathways for potential applications. Since endohedral metallofullerenes are stabilized through a charge transfer from the metal to the cage, the quantum yield of the erbium cation emission depends on the geometry of the cage, as was proven by the Ito et al. [4b].

Up to this point, we focused our discussion on zero-dimensional (0D), spherical nanoparticles as light-emitting materials. Recently a substantial research interest emerged for 2D materials due to their unique structural, electronic, and morphological properties. They can even be used as counterparts in molecular motors. While numerous 0D nanoparticles have been used as components in molecular motors, the first report of two-dimensional (2D) materials in such systems was presented in 2017 by Dong et al. [5a]. They used tetraphenylethylene building blocks to build 2D networks that could be used as sensors toward metal cations. The authors investigated the performance of the sensor against a series of cations, including transition metal ions (Ag^{+}, Mn^{2+}, Co^{2+}, Ni^{2+}, Cu^{2+}, Zn^{2+}, Cd^{2+}, Pb^{2+}, Al^{3+}, Ga^{3+}, Fe^{3+}, Ln^{3+}, Cr^{3+}), alkali metals (Li^{+}, Na^{+}, K^{+}), alkaline earth ions (Mg^{2+}, Ca^{2+}, Ba^{2+}, Sr^{2+}), and rare earth metal ions (Ce^{3+}, Eu^{3+}, Tb^{3+}, Er^{3+}). Among these metal ions, Cu^{2+}, Co^{2+}, Mn^{2+}, and Fe^{3+} exhibit strong quenching effects. In other examples, exfoliated, ultrathin, 2D ternary chalcogenide nanosheets have been used for DNA sensing based on their selective fluorescence properties. The 2D ternary chalcogenide nanosheets originated from flakes of Ta_2NiS_5 and Ta_2NiSe_5 [5b]. Dong et al. [5a] reported the first work on molecular motors with a 2D geometry. The terphenylene units were coupled through Sonogashira coupling by utilizing tetrakis (triphenylphosphine) palladium and copper(I) iodide in *N,N'*-dimethylformamide/triethylamine. The excitation maximum under which the materials were studied was 365 nm and they were tested toward organic solvent vapors and metal cations for sensing applications.

In this chapter, the focus will be on selective systems that have been recently developed through liquid–liquid interfaces, air–water interfaces as also additional information of relevant 2D systems. The classes of molecules that will be highlighted are including the perfluorinated polyanilines (PANIs), covalent organic frameworks, and MOFs. Further insights on the effect of the substitution position for perfluorinated PANIs on the physicochemical properties of the polymers will be outlined, while a short introduction on the fundamentals of photoluminescence is presented in following section.

5.2 Fundamentals of photoluminescence

5.2.1 Light emission

The photoluminescence phenomenon arises from are the emission of light from a substance after absorbance of photons and a subsequent interaction and energy exchange of the molecule with the surrounding environment. Two distinct processes that involve the emission of photons after excitation are fluorescence and phosphorescence, and they represent a light emission that arises from two different pathways. The name of the former originates from the Greek word phosphor. These two types of photoluminescence are distinguished based on the following rule: the fluorescence is a fast process from a singlet to a single state that takes place in the course of nanoseconds and the phosphorescence is a slow process from a triplet to singlet state with a significantly longer lifetime. In this chapter the discussion will focus mainly on the fluorescence properties of polymers, nanomaterials, and MOFs. It is worthwhile to present a short mention of the singlet oxygen generation and the subsequent fingerprint phosphorescence. In general phosphorescence is always red-shifted, narrower, and a much weaker process compared with fluorescence.

The short-lived fluorescence phenomenon was first observed on quinine solutions by Sir J. F. W. Herschel in 1845 and the term was coined by G. G. Stoke. Fig. 5.2 presents a schematic representation of the Herschel experiment, who observed that the otherwise colorless quinine molecule emits blue light when it is illuminated with filtered white light. In studies that followed shortly, George Stokes elaborated on the phenomenon in greater detail in 1852. Fig. 5.2 also presents the transitions that lead to light absorption, a singlet–singlet fluorescence, and a triplet–singlet phosphorescence (B) as also the chemical structure of quinine (C).

The emission maxima of the fluorescence and the quantum yield are depending on a series of factors both on the molecular level and the environment that surrounds the fluorophore, as also on intermolecular and intramolecular interactions. The initial process is that when the molecule is excited by a light source, electrons are migrating to excited electronic states. The electronic states of an organic molecule are composed of a

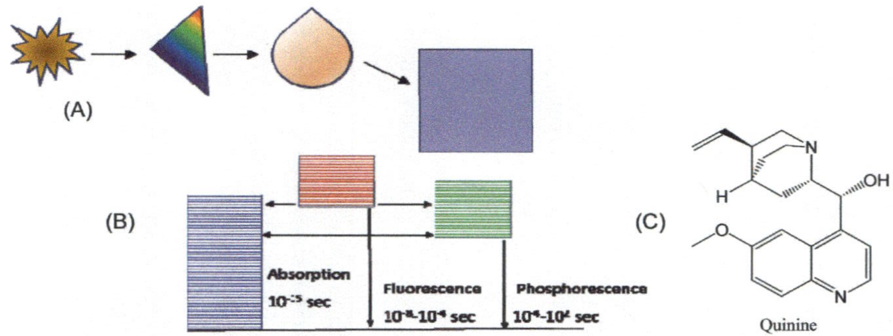

Figure 5.2 (A) In the 19th century, Herschel and Stokes after experiments with quinine tartrate salts employed salts due to low solubility of quinine in water, identified and coined the term fluorescence. (B) The transitions leading to fluorescence and phosphorescence (blue: S_0, red: S_1, green: T_1). (C) The chemical structure of quinine.

series of vibrational and rotational energy levels, all of them associated with the atomic nuclei orbitals. In the ground state all electrons are spin-paired, that is, they have opposite spin.

The light absorption and the subsequent excitation and emission are taking place between closely spaced vibrational and rotational energy levels. The absorption of light is a very fast process, in the time scale of femtoseconds. The light absorption, excitation, and emission are measured in quanta and the energy of each quantum according to Planck's Law is calculated by the equation: $E = hn = hc/l$ where E is the energy, h is Planck's constant, n and l are the frequency and wavelength of the incoming photon, and c is the speed of light.

The absorption of photons occurs with the light of specific wavelengths and hence represents specific interactions between the oscillating electric field of the photons with the charged electrons of the fluorophores. Hence the electronic absorption spectra are always broad due to the closely spaced vibrational levels. In classical physics terms the excitation of a molecule through absorption is taking place without a change in electron spin-pairing. To that end, the resulting excited state is a singlet, exactly as it is the ground state, the molecule will stay in the excited state. The excess of vibrational energy is converted into heat, which is then absorbed by neighboring solvent molecules upon colliding with the excited state fluorophore.

After absorption of photons a relaxation to the lowest vibrational energy level of the first excited state occurs, which is the S1 follows. This internal conversion (i.e., vibrational relaxation) represents a loss of energy in the absence of light emission. Photobleaching may occur in certain dyes, however in the majority of cases, the molecule will have to be excited hundreds to thousands of times before the highly reactive excited state molecule will show photobleaching.

5.2.2 Intersystem crossing

There are various ways that the energy of the excited state can be dissipated following the absorption of a photon. This can happen as a nonradiative process, for example, through dissipation as heat or the excited fluorophore will collide with a neighboring molecule and transfer energy to the later. This can lead to quenching or to the intersystem crossing where the molecule is now transferred to the lowest excited triplet state which is a classically forbidden state.

The intersystem crossing that involves a transition from the excited singlet S1 to the excited triplet T1 state is classically forbidden and rarely takes place. When it does take place though, two options are available: the fluorophore will decay through a phosphorescence with a long-lived emission or an energy transfer to nearby oxygen resulting in reactive oxygen species. The probability of the intersystem crossing is low due to the spin conversion and it is a very reactive excited state that may lead to photobleaching and the production of damaging free radicals. The formation of the reactive singlet oxygen and its detection and reactivity will be discussed in the following section.

5.2.3 Singlet oxygen generation

In other cases the intersystem crossing from the singlet to the triplet state is accompanied by the energy transfer to nearby oxygen molecules leading to excitation and the formation of the very reactive singlet oxygen [6]. After the excitation of the oxygen molecule to the singlet state, the relaxation to the ground state is accompanied by emitting light at a characteristic wavelength with $\lambda_{max} = 1270$ nm. This emission represents a fingerprint phosphorescence that has been repeatedly used for the detection of the singlet oxygen. Fig. 5.3 depicts the formation of the singlet from the triplet oxygen state after light irradiation. As we can see the oxygen has a total spin quantum number $S = 0$ and a multiplicity of state of 1. The energy of the singlet oxygen is 22.5 kcal higher than the ground state of the triplet oxygen lifetime of singlet oxygen is from 50 to 700 μs, depending on the solvent where the oxygen and the photosensitizer are dissolved.

Among the most efficient photosensitizers for the formation of the singlet oxygen are the small closed cage carbon nanomaterials, the fullerenes. Their intersystem crossing leads to a very low fluorescence quantum yield, at the level of 10^{-4} due to the small number of photons that return to the S_0 state. This singlet oxygen excited states are very reactive, and they represent a very electrophilic molecule. This excited oxygen state has low activation energy compared to the triplet state. Consequently it has a reaction rate with food products much greater than the ground, triplet, state. This has negative effects on food quality and storage. However, various organic compounds such as carotenoids and tocopherols in foods can minimize the oxidation through $^1O_2$ quenching. Fig. 5.4 presents the optical transitions that lead to the formation of the singlet oxygen and an

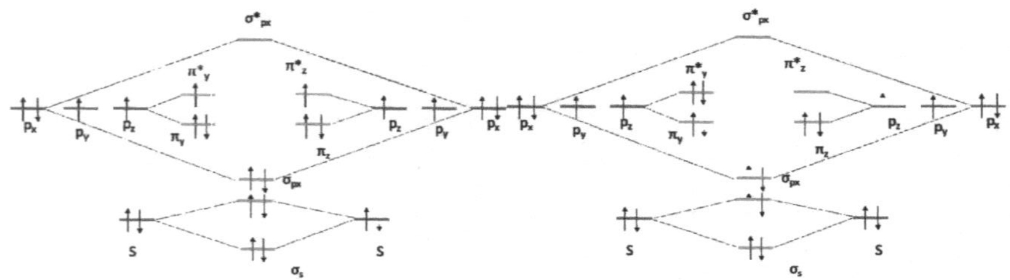

Figure 5.3 A schematic representation of the ground triplet (left) and the excited (singlet) oxygen formulations. The molecular orbitals are presented with green color and the atomic orbitals with red.

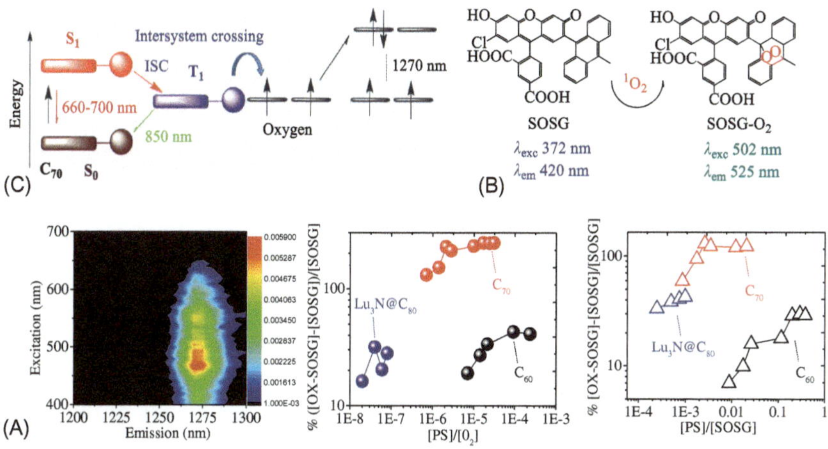

Figure 5.4 (A) The singlet–singlet and singlet–triplet optical transitions that lead to the singlet oxygen sensitization by fullerenes and a corresponding excitation-dependent photoluminescence map depicting the characteristic $\lambda_{max} = 1270$ nm emission (B). Reproduced with permission from Dallas et al. © John Wiley & Sons. (C) A sensor, the singlet oxygen sensor green has been used as a chemiluminescent sensor for the detection of the photosensitization properties of functionalized fullerenes. *Reproduced with permission from P. Dallas, S. Zhou, S. Cornes, H. Niwa, Y. Nakanishi, Y. Kino, et al., CF2 bridged C60 dimers and their optical transition, Chem. Phys. Chem. 18 (2017) 3540. P. Dallas, P.Q. Velasco, M. Lebedeva, K. Porfyrakis, Chem. Phys. Lett. 730 (2019) 130.*

excitation-dependent photoluminescence map for a unique $-CF_2$-bridged fullerene dyad synthesized through arc discharge where Teflon tubes are added inside the reactor [7]. Detailed studies of the empty cage and functionalized fullerenes demonstrated that the optimum wavelength for the formation of the singlet oxygen depends on the surface functionalization, that is, the number of adducts and their molecular structure.

5.2.4 Fluorescence microscopy: applications at the oil–water interface

Fluorescence microscopy is a technique that utilizes light emission by materials and molecules to unveil their structure and morphology. This technique has a sensitivity and information gathering ability that makes it an excellent candidate for elucidating the structure and dynamics of polymers and nanomaterials that are adsorbed at liquid–liquid interfaces.

It is feasible to monitor the liquid–liquid interface by utilizing the fluorescent light with the total internal reflection geometry, confocal imaging techniques, or multiphoton excitation techniques such as two-photon fluorescence microscopy [8]. For the case of two-photon excitation, it is restricted to a small region that is surrounding the focal plane of the laser beam. The two-photon fluorescence microscopy is a technique that has been used to unveil the polymer absorption at the toluene–water interface in a work by Rao et al. [8]. In two-photon fluorescence, the molecule absorbs two photons of a wavelength 2λ to be excited at a wavelength λ. So, in this example, two photons of 800 nm laser are used to excite at 400 nm (Fig. 5.5). The authors utilized the set up depicted in Fig. 5.5. They also found that the saturated surface excess was 1 mg/m^2 and corresponded to an area of 82 nm^2 per molecule. Since most polymers do not emit light, they used polystyrene, which is nonfluorescent, but labeled with fluorescein isothiocyanate. The molecular weight of the polystyrene used was 50,500 g/mol. In this system the fluorescein isothiocyanate that is used as a dye has an affinity for an aqueous phase; hence it can

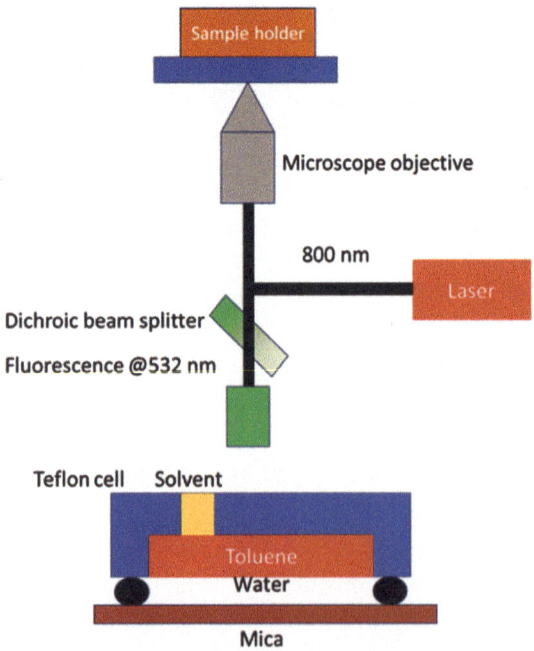

Figure 5.5 A schematic representation of the two-photon fluorescence system for the detection of the absorption of fluorescent polymers at the liquid–liquid interface, in this example, toluene–water. This figure is based on the work by Rao et al. *A. Rao, A. Dhinojwala, Two-photon fluorescence technique to study polymer adsorption at the toluene-water interface, Langmuir 19 (2003) 8813-8817.*

mitigate at the interface. Interestingly previous experiments with nonlabeled polymers did not show absorption at the oil–water interface.

Compared with single-photon confocal microscopy the two-photon fluorescence microscopy presents some advantages. First the photobleaching of the dye molecules due to localization of the excitation volume is kept to a minimal level. Second the excitation beam can be separated from the emitted fluorescence, and the scattering from the excitation beam will be omitted, hence a good sensitivity can be obtained.

The absorption of polymers at the liquid–liquid interface influences many industrial processes and laboratory reactions. Just a few examples include food processing, oil recovery, separation processes, and colloidal chemistry [9]. The narrow, 2D area of the interface is a great platform to study the pharmacological activity of drugs and a good base for the simulation of complex biomembranes [10]. In summary the detection, mapping, and sample analysis based on the fluorescence of the samples at liquid–liquid interfaces is invaluable in a series of environmental applications, for example, to detect toxic and polluting analytes such as isocyanates and pesticides. However, the synthesis and the monitoring of processes at the interface presents some limitations, for example, the temperature can only be between −5°C and +100 °C and for air–liquid interfaces between 5°C and 50°C, otherwise the interface cannot be maintained.

5.3 The superhydrophobic conductive polymers

In chapter 4, Synthesis and Generation of Polymeric Materials from Interfaces, we have discussed extensively the synthesis, formation of nanostructures, electronic transport and magnetic properties of PANI, polypyrrole, poly(3,4-ethylenedioxythiophene) (denoted as PEDOT) and other conductive polymers. Polyaniline, the most widely used conductive polymer, exhibits weak fluorescence with low-quantum yields and very few examples in the literature exist regarding its light-emitting behavior. Furthermore, for many applications, that is, detection of organic compounds, as well as solid contacts in ion-selective electrodes and thin films for gas sensing, conductive polymers need to be hydrophobic. Conductive polymers, despite being practically insoluble in most solvents, can be very well dispersed in aqueous media and may even absorb humidity from the atmosphere. Below we will give examples of the importance of a superhydrophobic behavior, especially when combined with conductive and light-emitting properties.

An important class of sensors is the ion-selective electrodes (ISEs). They consist of a polymeric membrane where receptors, called ionophores, can be incorporated. All solid-state electrodes without the inner filling solution but with ion-to-electron transducing material have also been developed. Conductive polymers have been widely used as solid contacts in this class of sensors, with PEDOT to be the predominant example. However, since a water layer at the ISE membrane will affect the electrode potential, the development of water-repelling conductive polymers is highly anticipated. For example, the superhydrophobic properties of (2-*n*-tetradecyl-2,3-dihydrothieno-[3, 4-*b*] [1,4]-dioxine) and the PEDOT-C1 4 (tetrakis-(pentafluorophenyl)borate) are necessary to prevent the detachment of the ion-selective membrane from its solid contact as well as to stop any accumulation of an aqueous film between the ion-selective membrane and the solid contact [11]. Potential accumulation of an aqueous film between the membrane and the solid contact is something that has a detrimental effect on sensor performance. Fig. 5.6 presents the structure of a typical ion-selective electrode and the potential-time transients of three different conducting substrates, gold, platinum, and glassy carbon. The solid contact PEDOT (PSS −)-electrodes are potassium selective and have undergone conditioning in blood serum-like electrolyte solution (BSE-1). Three different electron-conducting substrates were studied, namely, Pt, Au, and GC [12]. A highly hydrophobic PEDOT derivative acts as an intermediate layer for solid-state ion-selective electrodes (for H + , K + , and Na +). It has excellent theoretical response slope, short equilibration time, and excellent potential stability [13].

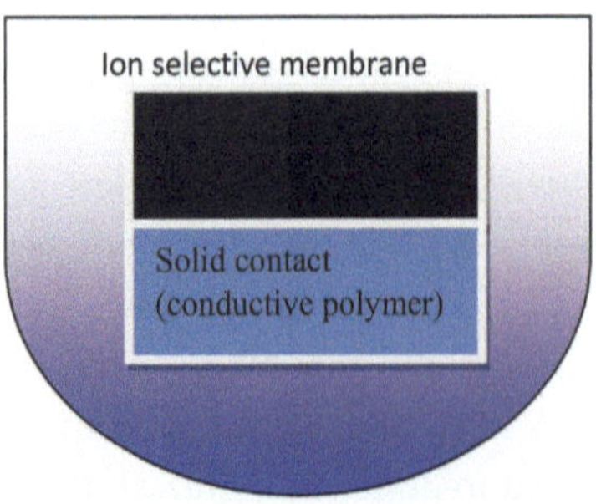

Figure 5.6 A schematic representation of an ion-selective electrode.

5.3.1 Linear 3-perfluoroctyl polyaniline

The effect of the interfacial tension (the force arising from an imbalance of forces in the interface) on the physical–chemical properties of the interface and on the self-assembly of materials at the interface has been outlined in Chapters 2, Nature of the Liquid–Liquid Interface, and 3, Self-Organisation of Nanomaterials at the Interface. These differences among various interfacial systems have been utilized for the synthesis of a series of fluorescent materials by Dallas and coworkers. They demonstrated that the different interfacial tension during oxidative polymerization of perfluorinated aniline results in control of the light-emitting properties of polymers [14]. In these seminal studies, five organic solvents were employed as the medium in which the monomer was dissolved. The interfacial tension for these systems exhibited the following values (in mN/m) against water: a perfluorinated carbon-like hydrofluoroether (3 M Novec 7300) ≈ 52; chloroform = 31.6; dichloromethane = 28.2; toluene = 36.1, 1,2-dichlorobenzene = 40.

The reaction proceeds in room temperature without any stirring and two different classes of materials can be isolated. An insoluble film that is formed at the interface and the soluble oligomers that exhibited strong fluorescence emission in the visible range. Fig. 5.7 demonstrates the polymerization procedure and representative scanning electron microscopy (SEM) images of the polymers isolated as a thin film from the interface. Interestingly the particles appear to have a "doughnut" shape, with 1–3 holes [15]. This number of holes in the polymer structure represents the so-called "genus" (g) of a polymer particle, and its definition comes from differential geometry. According to previous works very high genus numbers (≤ 100) have been observed in polymer systems.

The insoluble component in the reaction mixtures consists of polymer films or nanocomposites with various nanomaterials that are isolated from the interface. The films can be easily removed from the interface and subsequently washed with chloroform and water to remove any absorbed oligomers and remaining acids. These polymers were not fluorescent or exhibited weak emission with a very short Stokes shift. On the other hand the oligomers of these perfluorinated PANIs were proven to be highly fluorescent with complicated optical spectra. Fig. 5.8 demonstrates the excitation-dependent photoluminescence maps of a soluble oligomeric mixture of products derived from

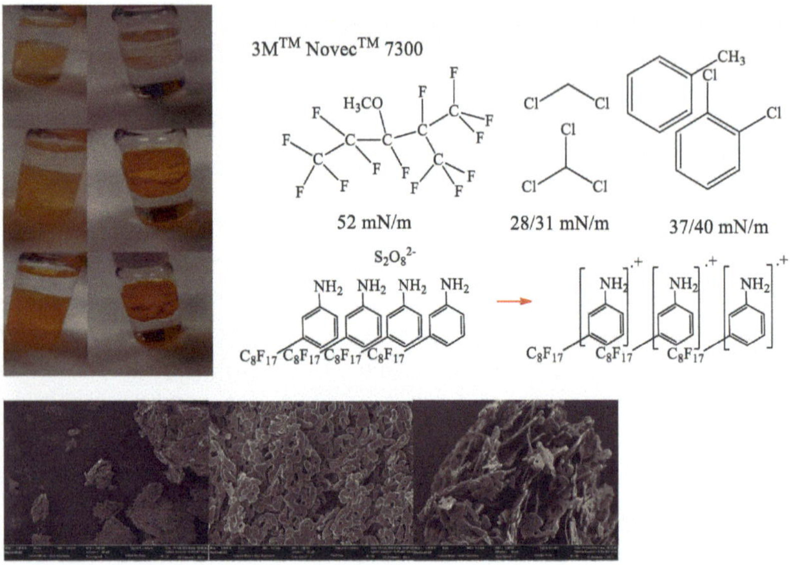

(A)

(B)

Figure 5.7 (A) The interfacial polymerization may proceed through different organic–water interfaces each one with different characteristics and interfacial tension. Typical interfacial tension values (in mN/m) for various solvent pairs are: (I) water–perfluoroether: (52); (II) water–chloroform: (31); (III) water–toluene: (37); (IV) water–dichloromethane: (28); and (V) water–*o*-dichlorobenzene: (40). SEM images of the insoluble fraction derived from (B) chloroform and the genus number based on their doughnut shape. *Adapted from P. Dallas, I. Rasovic and K. Porfyrakis, Mapping and tuning the fluorescence of perfluorinated polyanilines synthesized through liquid–liquid interfaces, J. Phys. Chem. B, 120 (2016) 3441, with permission by the American Chemical Society.*

chloroform–water interface. The different fractions were separated through injection in a silica column (see the corresponding images) and four products were isolated and characterized further using various methods. The oligomers were solubilized in toluene prior to separation and toluene was also the eluent phase. The elution for each of the four fractions is indicated next to the images of the elution of oligomers. Their photoluminescence maps demonstrated the existence of different radiative pathways, and the nuclear magnetic resonance spectroscopy measurements were employed to resolve the structure and correlate the different oxidation states and cross-linked structures with their optical properties.

Photoluminescence analysis of the emission and excitation of the four fractions (Fractions 1–4) indicated the following excitation (λ_{exc})

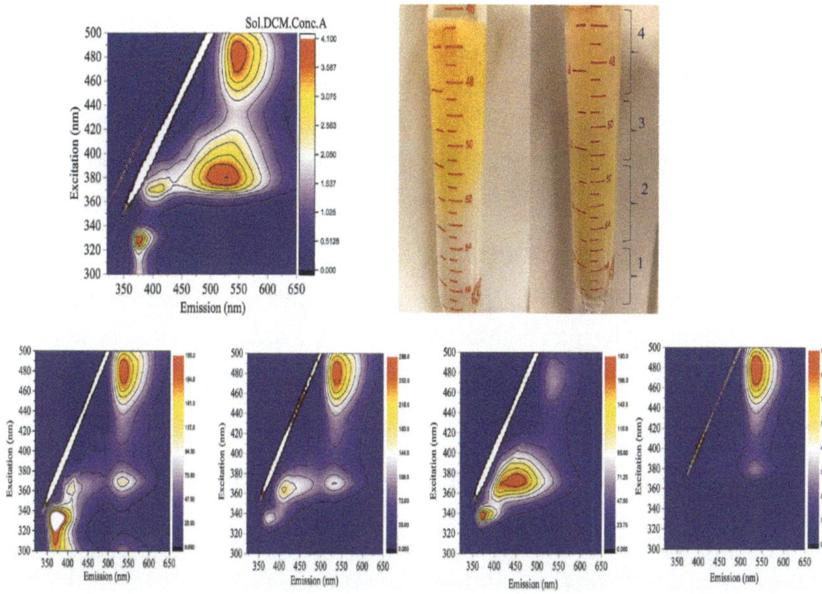

Figure 5.8 By passing soluble fraction through a silica column, it is possible to separate different soluble oligomeric fractions of light-emitting perfluorinated polyaniline. (A) Their corresponding photoluminescence excitation-dependent spectral maps recorded in toluene solutions are shown in (B). The materials were derived from chloroform and were redissolved in toluene before injection onto a silica column due to optimum interaction with silica. *Adapted from P. Dallas, I. Rasovic and K. Porfyrakis, Mapping and tuning the fluorescence of perfluorinated polyanilines synthesized through liquid–liquid interfaces, J. Phys. Chem. B, 120 (2016)3441. P. Dallas, I. Rasovic, T. Puchtler, R.A. Taylor, K. Porfyrakis, Long Stokes shifts and vibronic couplings in perfluorinated polyanilines, Chem. Comm. 53 (2017) 2602. Reproduced with permission from the American Chemical Society.*

and emission (λ_{em}) maxima: Fraction 1 with $\lambda_{exc} = 330$ nm; 370 nm; 480 nm and $\lambda_{em} = 368$ nm; 487 nm; 529 nm – Fraction 2 with $\lambda_{exc} = 335$ nm; 364 nm; 480 nm and $\lambda_{em} = 378$ nm; 410 nm; 532 nm – Fraction 3 with $\lambda_{exc} = 335$ nm; 364 nm; 480 nm and $\lambda_{em} = 378$ nm; 452 nm; 532 nm – Fraction 4 with $\lambda_{exc} = 373$ nm; 475 nm and $\lambda_{em} = 536$ nm. This work provided a pathway for the development of a new class of fluorescent labels and polymeric sensors through a low cost, easily scalable reaction at room temperature. It is expected that further improvements of the system will result in different oligomers and polymers with better quantum yields, longer Stokes shifts, as well as emission and excitation maxima that can be tuned at will. Since the purification of the fluorescent oligomers takes place through slow silica flash chromatography, it is important to find suitable functionalized columns for high-performance liquid chromatography. This would lead to significant time gains regarding the purification of the compounds and will enable the scaling-up of the oligomers production.

5.3.2 Branched 4-perfluoroctyl aniline

The second monomer of the perfluorooctyl aniline that will be discussed in this chapter is the para-substituted aniline, where the para position is blocked. The blocking of the para position by the perfluorocarbon groups is consequently forcing the polymer to crosslink and prevent the formation of a linear backbone. This difference in the backbone arrangement has profound differences in the optical and electrochemical properties of the materials.

The first report on the polymerization of the 4-perfluoroctyl aniline monomer was the study published in 2017 by Dallas et al. [14b]. The polymerization was conducted at the interface between water and an organic solvent with or without the presence of surfactants, namely, the Triton X-100 in the aqueous phase. Fig. 5.9 demonstrates a schematic representation of the polymerization process when micelles are formed in the aqueous phase, images of the two-phase systems with different molar ratios of reagents and the positions where substitution is preferred in the 3- and the 4-perfluorooctyl aniline. In the case of 4-perfluoroctyl aniline, the amine group is directing the substitution to an ortho or para position compared to it, as it is an electron-donating group. On the other hand, the electron-withdrawing perfluorocarbon groups are directing the substitution to a meta position relative to them as they represent a strongly deactivating group. To that end, we propose the substitution pattern given in Fig. 5.9C.

After purification of the products, the structure of the oligomers was identified through ^{1}H and ^{19}F NMR spectroscopy. The structures of two oligomers synthesized with two different amounts of Triton X-100 is depicted in Fig. 5.10. An exciting observation was the clear patterns of vibronic features on the excitation-dependent photoluminescence maps. We see three representative maps in Fig. 5.10B, where the aqueous phase contained 1 mg, 10 mg, and 100 mg Triton X-100, respectively. The surfactant, Triton X-100, was added to alter the characteristics of the interface and lead to different products with

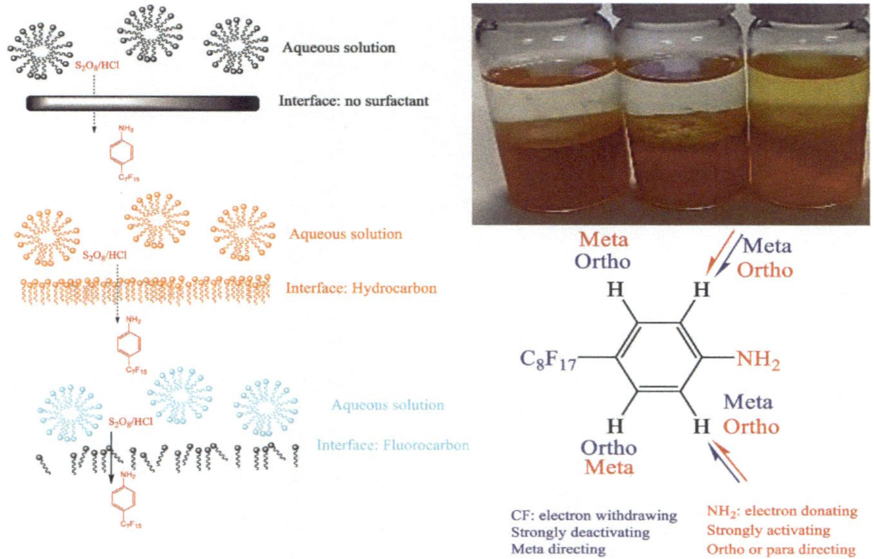

Figure 5.9 A schematic representation for the formation of micelles and the interface structure when no surfactant is added; when it is added in a hydrocarbon–water system and fluorocarbon–water system. The interfacial polymerization from an HFE–water system with the following amounts of monomer (*x*), acid (*y*), oxidant (*z*), surfactant (*s*) from left to right: (A) $x = 240$ mg, $y = 0.1$ mL, $z = 240$ mg, $s = 14$ mg; (B) $x = 240$ mg, $y = 0.3$ mL, $z = 720$ mg, $s = 50$ mg; (C) $x = 240$ mg, $y = 0.3$ mL, $z = 720$ mg, $s = 68$ mg. Notice the migration of greenish oligomers to the aqueous phase in (C). (C) Promoted positions of an electrophilic attack on 4-perfluoroctyl aniline where the para substitution position relative to the amine group is blocked.

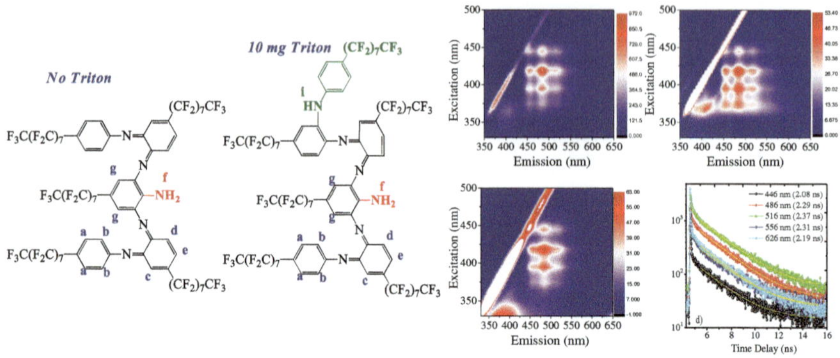

Figure 5.10 ^{1}H NMR data for samples derived from toluene without surfactant, and from toluene with the addition of 10 mg Triton (Fr.1). We assign two core oligomers, a fully oxidized pernigraniline, and a further crosslinked structure. The 2D COSY NMR supported the structure assigned in the ^{1}H NMR. The doublets of protons *a* and *b* are coupled in the same benzene ring, as are the doublets of protons d and e. Excitation-dependent photoluminescence maps for the toluene-derived first fractions, when synthesized with the following amounts of Triton X-100 (A) 1 mg (B) 10 mg (C) 100 mg. (D) Time-resolved fluorescence for the sample (B), Fraction 1. The lifetimes are indicated in parentheses. *Reproduced with permission from P. Dallas, I. Rasovic, T. Puchtler, R.A. Taylor, K. Porfyrakis, Long Stokes shifts and vibronic couplings in perfluorinated polyanilines, Chem. Comm. 53 (2017) 2602 the Royal Society of Chemistry.*

variable geometry and optical properties. The critical micelle concentration for Triton X-100 is 0.25 mM (0.162 mg/mL).

Hence the reaction took place at a concentration below (when toluene—water had 0.125 mg/mL of surfactant in the aqueous phase) and above the critical micelle concentration (when the aqueous phase had 1.25 and 12.5 mg/mL for toluene—water and 1.75 mg/mL for HFE—water system). The materials derived from toluene and HFE demonstrated the longest Stokes shift and unique vibronic features in the PL spectra, while from dichloromethane, chloroform, toluene, and hydrofluoroether—water systems in the absence of any surfactants demonstrate only a weak ultraviolet emission.

An interesting pattern was observed for the toluene—water/Triton interfacial polymerization reaction. Three different fractions through column chromatography were isolated from the reaction mixture. The first fraction that was eluted with toluene from low retention times exhibits a red-shifted pattern with pronounced vibronic features. These features are becoming significantly weaker and nearly disappear in the excitation-dependent photoluminescence mapping of the second fraction which was eluted at higher retention times. The quantum yields (QY) for typical materials are outlined in the following discussion. The quantum yield for the oligomer exhibiting the unique vibronic features from toluene solution was calculated to be 2.2% for the sample with vibronic features derived from HFE (Fr.1) it was 5%; while for the sample with the exceptionally long Stokes shift, derived also from HFE, the QY was significantly lower, 0.5%. We then proceeded in calculating the molar extinction coefficient by employing pyrene as a reference for the Fraction 1 from toluene. For the maximum absorbance located at the ultraviolet region, the extinction coefficient was 4536 and for the visible region absorbance (n−p* transition), it was 687 cm^{-1} M^{-1}.

Herein a more detailed discussion on the light fluorescent properties of the oligomers will be presented. While the first fraction derived from the hydrofluoroether, in the presence of Triton demonstrated a pattern with strong vibronic couplings, similar to toluene-derived sample separated by 1281 and 1414 cm^{-1} for the second fraction, unusually high Stokes shift of more than 200 nm were observed. Furthermore while only one peak for terminal $-NH_2$ groups can be seen in the toluene-derived samples, the HFE derivative exhibits three terminal amine peaks, a further indication that an $-NH_2$ group will further redshift the emission. These findings are in good agreement with the MALDI-TOF data that were recorded with indicating increased cross-linking for the HFE derivatives.

The fluorescent color of the different fractions under 400 nm laser excitation is shown in Fig. 5.11. The first eluted fraction with toluene as the eluent phase is depicting a greenish color, while the second fraction that is eluted in higher retention times is showing a yellowish color. The excitation dependence photoluminescence maps are showing the absence of vibronic couplings in the case of the oligomers with long Stokes shifts. Fig. 5.11c displays the UV−visible spectra of the three different fractions. The first two, fluorescent, fractions exhibit a broad and strong absorbance peak in the range of 360 to 600 nm which is assigned to the n−p* electron transition. Fig. 5.11d presents the time-resolved fluorescence analysis data along with the corresponding fittings for a mono or a biexponential decay.

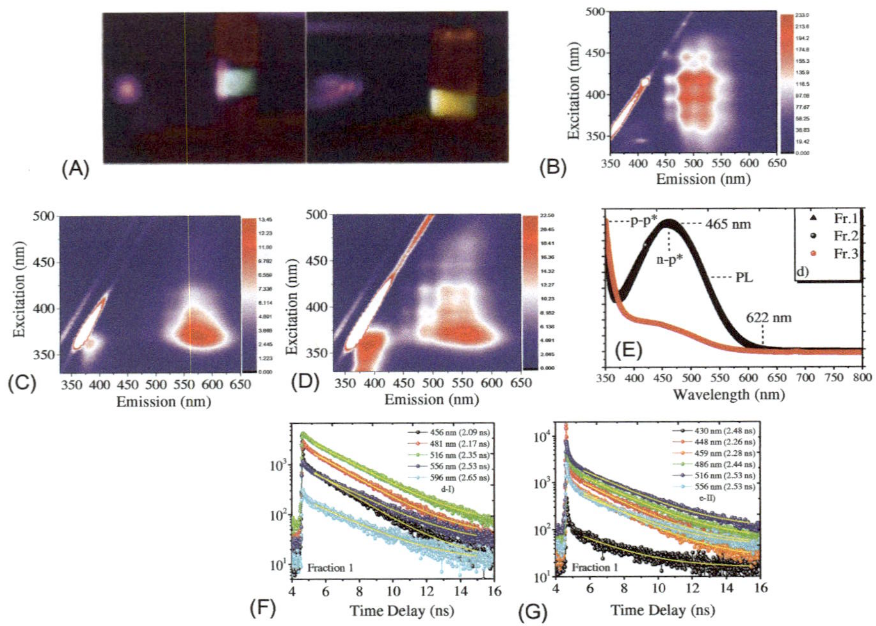

Figure 5.11 (A) Images under 400 nm laser irradiation for two different oligomer fractions as eluded by the silica column. Fraction 1 appears as greenish and fraction 2 as yellowish. (B–D) Excitation-dependent photoluminescence maps and the UV–vis spectra (E) for the three separated fractions of the HFE–water–Triton system. The strong band at 465 nm is associated with conjugated aromatic amine polymers (n–p* transition). (F–G) Time-resolved fluorescence for the first two fractions. *Reproduced with permission from P. Dallas, I. Rasovic, T. Puchtler, R.A. Taylor, K. Porfyrakis, Long Stokes shifts and vibronic couplings in perfluorinated polyanilines, Chem. Comm. 53 (2017) 2602, the Royal Society of Chemistry.*

5.3.3 Effect of substitution position on the physical properties. Cross-linked versus branched chains

The most common medium for the polymerization of aniline in the aqueous solution. In this case the oxidation of the aniline monomer in water, the growth of the chain is passing through various stages as proven by the work of the Stejskal group [16a]. According to their findings, when the polymerization was conducted in the absence of any added acid it is going through two distinct exothermic phases. βy carefully collecting and studying the intermediates of the reaction, the mechanism of the oxidative polymerization was revealed, including the growth of the PANI nanotubes.

During the first stages of the reaction, the formation of oligomers is observed. The second step involves the formation of crystals that will subsequently lead to nanotubes growing from them as proven by optical spectroscopy studies of the intermediates of the reaction.

In this case, the precipitated oligomers acted as a template with nanotubes growing while the oligomers acted as seeds. Scanning electron microscopy revealed the formation of nanotubes with 200 nm diameter as other morphologies, for example, nanorods, nanobubbles, and nanosheets. Phenazine units have also been observed through spectroscopic studies [16a]. These studies have revealed the complexities of the oxidative polymerization of PANI even in the simplest, linear form. As we discussed in the previous paragraphs, the two different substitution positions lead to a significant change on the polymer geometry: while 3-perfluoroctyl PANI grows in a linear geometry, the para-substituted 4-perfluoroctyl aniline due to the blocking of the para position grows in a zigzag backbone and this appears to have a significant influence on the optical properties.

In previous works different conventional PANI backbones were tested as an evaluation of the effect of the polymer arrangement. The comparison was based on capacitance measurements and demonstrated that the linear PANI possessed a significantly higher capacitance. However, an interesting feature is that after subsequent cycles, the branched PANI increases its capacitance for orders of magnitude. To be more precise, it increased from 4 to 294 F/g after the 107 charge–discharge cycles and subsequently decreased slowly to 180 F/g after 1000 charge–discharge cycles due to the formation of phenazine units [16b]. In this seminal work the authors demonstrated the novel capacitance properties of PANI with branch-like chains. This is a unique capacitance performance for PANI and the effect of the linearity or not of the macromolecular chains was investigated through extensive electrochemical measurements. The PANI polymer that exhibits the branch-like molecular geometry was studied by using SEM and Fourier transform infrared spectroscopy (FTIR). Regarding the morphology of the polymers, the branch-like polymer grows in a rhombic straw mat-like morphology, while the common PANI is forming nanoparticles and the branched PANI ruptures into small linear oligomeric fragments that lead to a higher $\pi-\pi$ molecule conjugation in PANI chains and a better electrical conductivity as well as a lower IR drop.

5.4 Other photofunctional 2D materials

During the last decade an enormous effort was dedicated to the development of 2D materials, fueled by the discovery and the exciting properties of graphene. Various 2D light-emitting materials such as transition-metal dichalcogenides (TMDs) and phosphorene have been extensively studied and regarding their optical and optoelectronic properties that are dictated by the control of the number of layers, dielectric environment, and formation of heterostructures. This class of materials includes 2D organic and organic–inorganic hybrid materials, polymers, MOFs, and perovskites. 2D supramolecular polymers have also found applications in the fields of photoelectric sensing, information storage, and fluorescence enhancement [17]. To ensure that a 2D growth is taking place without 3D growth, the counterparts are restricted in a 2D space.

Graphene, boron nitride, and various chalcogenides (molybdenum disulfide, tungsten sulfide) are the most important 2D nanosheets and have triggered extensive research activity. These materials are, in most cases, synthesized through chemical vapor deposition (CVD) methodologies using metal catalysts. In case of graphene, exfoliation with liquid chemistry methods is an effective means for the preparation of individual nanolayers. These 2D materials have found applications as components of electronic devices as well as in photonics and spintronics applications [18]. Besides the classes of materials that are synthesized and grown through CVD methods, other types of 2D nanosheets are synthesized through other bottom–up approaches by utilizing ionic and molecular components [19]. Examples of those methodologies include 2D anthracene-based photopolymerized materials assembled at an air–water interface, and cycloaddition-grown anthracene nanosheets with carbon–carbon covalent bonds.

5.4.1 Metal–organic framework nanosheets through liquid–liquid or air–liquid interfaces

Metal–organic frameworks are a versatile class of materials with different applications with respect to their design flexibility and physical properties that can be selected at will [20].

An interesting aspect of the MOFs is that they can exhibit solid-state reactivity, as shown by Skoulika and coworkers who, after thermal treatment, demonstrated cross-linking among sp-hybridized triple bonds in a cadmium–acetylacetonate complex [21a]. Sakamoto and coworkers expanded these concepts further, through the synthesis and detailed study of a *bis*(dipyrrinato)zinc(II) complex nanosheet that utilized a dipyrrin ligand and zinc(II) cations [22]. The authors achieved a large domain size (>10 μm), guest inclusion, a stepwise layer formation method, and photoelectric conversion of the material. It was proven that the liquid–liquid interface is essential for the synthesis of the film. The authors performed the synthesis in a one-phase system as well, for example, in dichloromethane at room temperature or *n*,*n*-dimethylformamide at 105°C and both approaches resulted in a powder product. While the liquid–liquid interfacial synthesis resulted in a multilayer nanosheet, on the other hand, when the complexation was conducted at an air–liquid interfacial reaction it produces a single layer with domain sizes of 410 mm on one side. The bis(dipyrrinato)zinc(II) metal complex nanosheet was deposited on various substrates using the Langmuir–Schafer process. When used as a photoanode in a photoelectric conversion system the complex was the first photofunctional bottom–up nanosheet. Other works also explored the photophysical properties of bis-(dipyrrinato)–metal complexes to understand and evaluate their potential as antenna pigments in light-harvesting devices and arrays. The bis(dipyrrinato)–metal complexes that were initially synthesized by Fischer in 1924 has been considered nonfluorescent by most scientists. However, it has recently been shown that simply through a replacement of the phenyl rings at the 5,5′-positions of bis(dipyrrinato)–zinc complex with mesityl groups radical changes are caused. This substitution changes the complex, and from a weak emitter that deactivates within ~100 ps of photoexcitation, it is now a highly fluorescent chromophore with a multinanosecond singlet excited-state lifetime [23]. The bis-(dipyrrinato)–metal complexes can be used as self-assembling antenna pigments and components in synthetic light-harvesting architectures.

The bis(dipyrrinato)—metal complexes, were usually considered non-fluorescent but by replacing the phenyl rings at the 5,5′-positions of bis (dipyrrinato)—zinc complex with mesityl groups this picture changes. From a weak emitter with a fluorescence lifetime of 100 ps, it transforms into a strong emitter with an emission lifetime in terms of several nanoseconds. Both liquid—liquid and air—liquid interfaces have been used as mentioned in the comprehensive review by Sakamoto et al. [24].

In the case of hexafunctional terpyridine monomers that can be laterally connected by metal salts, the nanolayers can be transferred from the air—water interface to a solid substrate with a thickness of 1.4 nm [25]. These methodologies gave further insights into the synthesis of 2D nanosheets, with the liquid exfoliation method being the most prominent strategy for it. Coleman and coworkers detailed the liquid exfoliation of molybdenum disulfide (MoS_2), tungsten disulfide (WS_2), molybdenum diselenide ($MoSe_2$), molybdenum ditelluride ($MoTe_2$), tantalum diselenide ($TaSe_2$), niobium diselenide ($NbSe_2$), nickel diteluride ($NiTe_2$), boron nitride (BN), and bismuth telluride (Bi_2Te_3) [26].

5.4.2 Covalent organic frameworks

Covalent organic networks (COFs) are all-organic analogs of the MOFs. In this case, the covalent bonds are those directing the self-assembly and the material might expand in 2D or 3D depending on the torsion angle of the bridging units [27].

Covalent organic frameworks represent a versatile tool for the materials chemist due to the countless possible combination of ligands and central organic cores that can lead to a plethora of materials. Van Der Waals interaction dictates the layer-by-layer assembly hence the intralayer interactions are giving rise to crystalline domains while the weak nature of the bonding enables a layer exfoliation toward single-layer materials. This class of materials due to the porosity and functional groups on the polymeric backbone can be used for gas storage, gas separation, and synthesis of functional composites.

As it is common for layered materials, for example, graphene and chalcogenides, the exfoliation as also the reassembly can have a pronounced effect on the electrochemical, electronic transport, and optical properties of the materials. A seminal work by Mal et al. demonstrated that the reassembly of 2D layers in a selective and responsive mode results in characteristic changes on the fluorescence. The covalent networks with the ionic character that they synthesized, exhibited an orange fluorescence upon coupling with DNA. Extensive time-resolved and steady-state fluorescent measurements revealed that the 2D COFs are capable of selective double-stranded DNA compared with single-stranded DNA. This was proven by selected area electron diffraction resulted in a crystalline hybrid material [28].

Schrettl et al. [21b] created a carbon-rich 2D material using carbonization methods. They used amphiphiles that contain hexayne segments [21c] as metastable carbon precursors and self-assembled these into ordered monolayers at the air–water interface. Subsequent carbonization by ultraviolet irradiation in ambient conditions resulted in the quantitative carbonization of the hexayne sublayer.

In a very well-presented perspective, Wang et al. highlighted the main procedures and characteristics of covalent organic frameworks through liquid–liquid interfaces [29]. On the other hand, the air–water interface can also be used for the synthesis of such networks, for example, from dimerization of anthracene-based monomers [30]. While more unique interfacial systems such as the dichloromethane/trifluoromethanesulfonic acid (CH_2Cl_2–CF_3SO_3H) and the benzene/trifluoromethanesulfonic acid have also been used for the synthesis of 2D materials, for example, through cyclotrimerization reactions. Liu et al. synthesized single-layer as also few-layer triazine-based networks via the trimerization of carbonitrile at the interface of dichloromethane and trifluoromethanesulfonic acid, one of the strongest acids that acted as a catalyst. The obtained materials were crystalline and were applied as the active semiconducting layer in a field-effect transistor with a bipolar behavior with a high on/off ratio of 10^3 and electron mobility of 0.15 $cm^2/V\ s$ [31]. The authors have proven that this triazine-based 2D network demonstrated the highest mobility among other competitive materials, making it a very strong candidate for applications in electronic devices. TEM images revealed the striking difference between the triazine-based nanoparticles synthesized in pure CF_3SO_3H and the layers from CH_2Cl_2–CF_3SO_3H. The yield was 68% and under UV light irradiation, the triazine-based nanosheets demonstrated a bright blue–green fluorescence.

5.5 Conclusions

This chapter discussed in detail the formation and the light-emitting properties of a series of polymers and 2D materials synthesized through liquid–liquid interfaces. Emphasis was given on the perfluorinated PANI s, a class of materials that have been recently developed, due to their ease of preparation, low cost, tunable fluorescent properties, and straightforward purification. These oligomers have been already used as sensors toward the detection of small organic molecules, both acids and bases. Conventional PANI is an important conductive material with a slightly hydrophilic character and very high conductivity values. However, the affinity with water might hinder potential applications and the fluorescence of the polymer is very weak with low-quantum yield and strong light absorption throughout the visible range. For example, for their use as solid contacts in ion-selective electrodes, the materials need to be hydrophobic, thus preventing the accumulation of an aqueous film between the membrane and the solid contact which will hinder the performance of the cell. Interestingly the perfluorinated derivatives exhibit a relatively strong fluorescence that can be tuned at will, for example, under the presence of organic acids or bases and when the polymer is synthesized at different backbone configurations. Further development of the materials, that is, synthesizing thin films with high capacitance and conductivity will pave the way for applying them as electrochemically active water-repelling coatings. The interfacial tension of the interface can dictate the reaction yield and the different products, while the addition of polymers and surfactants can alter the mechanical properties of the films. Finally their contact angle with water can be high enough to consider them superhydrophobic ($>140°C$). The MOFs and the covalent frameworks are other important systems that can be synthesized through interfaces and possess photophysical and electronic transport properties. The main challenge is their efficient transfer to substrates due to their potentially poor mechanical properties. However, 2D networks have been used as semiconductor components in electronic devices and the i–v curves have been recorded while demonstrating excellent mobility.

References

[1] M.J. Krysmann, A. Kelarakis, P. Dallas, E.P. Giannelis, Formation mechanism of carbogenic nanoparticles with dual photoluminescence emission, J. Am. Chem. Soc. 134 (2012) 747.
[2] K.P. Carter, A.M. Young, A.E. Palmer, Fluorescent sensors for measuring metal ions in living systems, Chem. Rev. 114 (2014) 4564–4601.
[3] a) M. Kanj, H. Rashid, E.P. Giannelis. 2011, Industrial first field trial of reservoir nanoagents, Society of Petroleum Engineers, Middle East Oil and Gas Show and Conference in Manama, SPE (142592).

b) Z. Hu, H. Gao, S.B. Ramisetti, J. Zhao, E. Nourafkan, P.W.J. Glover, et al., Carbon quantum dots with tracer-like breakthrough ability for reservoir characterization, Sci. Total Environ 669 (2019) 579.
c) W. Yang, H. Zhang, J. Lai, X. Peng, Y. Hu, W. Gu, et al., Carbon dots with red-shifted photoluminescence by fluorine doping for optical bio-imaging, Carbon 128 (2018) 78.
[4] a) V.K. Singh, A. Singh, V.K. Gupta, R.A. Singh, Radical cyclization of arenesulfonyl chlorides and alkynes: a rapid access to π-conjugated benzothiohenes, Int. J. Org. Chem. 6 (2016) 55.
b) Y. Ito, T. Okazaki, S. Okubo, M. Akachi, Y. Ohno, T. Mizutani, et al., Enhanced 1520 nm photoluminescence from Er^{3+} Ions in di-erbium-carbide metallofullerenes $(Er_2C_2)@C_{82}$ (isomers I, II, and III), ACS Nano. 1 (2007) 456.
[5] a) J. Dong, K. Zhang, X. Li, Y. Qian, H. Zhu, D. Yuan, et al., Ultrathin two-dimensional porous organic nanosheets with molecular rotors for chemical sensing, Nat. Commun 8 (2017) 1142.
b) Z. Zhenxi, Facile one-pot synthesis of multifunctional polyphosphazene nanoparticles as multifunctional platform for tumor imaging, Anal. Bioanal. Chem. 410 (2018) 3723–3730.
c) C. Tan, P. Yu, Y. Hu, J. Chen, Y. Huang, Y. Cai, et al., High-yield exfoliation of ultrathin two-dimensional ternary chalcogenide nanosheets for highly sensitive and selective fluorescence DNA sensors, J. Am. Chem. Soc 137 (2015) 10430.
[6] a) P. Dallas, G. Rogers, T. Puchtler, R.A. Taylor, G.A.D. Briggs, H. Shinohara, et al., Charge separated states and singlet oxygen generation of mono and bis adducts of C_{60} and C_{70}, Chem. Phys. 465 (2016) 28.
b) D.B. Min, J.M. Boff, Chemistry and reaction of singlet oxygen in foods, Compr. Rev. Food Sci. Food Safety 1 (2002) 58.
[7] a) P. Dallas, S. Zhou, S. Cornes, H. Niwa, Y. Nakanishi, Y. Kino, et al., CF_2 bridged C_{60} dimers and their optical transition, Chem. Phys. Chem 18 (2017) 3540.
b) P. Dallas, P.Q. Velasco, M. Lebedeva, K. Porfyrakis, Chem. Phys. Lett. 730 (2019) 130.
[8] A. Rao, A. Dhinojwala, Two-photon fluorescence technique to study polymer adsorption at the toluene-water interface, Langmuir 19 (2003) 8813–8817.
[9] A.G. Volkov, D.W. Deamer (Eds.), Liquid-Liquid Interfaces: Theory and Methods, CRC Press, Boca Raton, FL, 1996.
[10] A.W. Adamson, Physical Chemistry of Surfaces, Wiley-Interscience, New York, 1990.
[11] M. Guzinski, J.M. Jarvis, P. D'Orazio, A. Izadyar, B.D. Pendley, E. Lindner, Solid-contact pH sensor without CO_2 interference with a superhydrophobic PEDOT-C14 as solid contact: the ultimate water layer, Anal. Chem. 89 (2017) 8468.
[12] a) M. Guzinski, J.M. Jarvis, B.D. Pendley, E. Lindner, Equilibration time of solid contact ion-selective electrodes, Anal. Chem. 87 (13) (2015) 6654–6659.
b) E. Zdrachek, E. Bakker, Electrochemically switchable polymeric membrane ion-selective electrodes, Anal. Chem 90 (2018) 7591–7599.
[13] M. Guzinski, J.M. Jarvis, B.D. Pendley, E. Lindner, Equilibration time of solid contact ion-selective electrodes, Anal. Chem. 87 (2015) 6654–6659.
[14] a) P. Dallas, I. Rasovic, K. Porfyrakis, Mapping and tuning the fluorescence of perfluorinated polyanilines synthesized through liquid–liquid interfaces, J. Phys. Chem. B 120 (2016) 3441.
b) P. Dallas, I. Rasovic, T. Puchtler, R.A. Taylor, K. Porfyrakis, Long Stokes shifts and vibronic couplings in perfluorinated polyanilines, Chem. Comm. 53 (2017) 2602.
[15] C.K. Haluska, W.T. Gozdz, H.G. Dobereiner, S. Forster, G. Gompper, Giant hexagonal superstructures in diblock-copolymer membranes, Phys. Rev. Lett. 89 (2002) 238302.

[16] a) M. Trchovå, I. Šeděnková, E.N. Konyushenko, J. Stejskal, P. Holler, G. Ćirić-Marjanovi, Evolution of polyaniline nanotubes: the oxidation of aniline in water, J. Phys. Chem. B 110 (2006) 9461.
b) J. Sun, L. Chen, Novel capacitance properties with branch-like chains, React. Funct. Polym. 138 (2019) 55–9461.
[17] a) Z. Wang, Q. Jingjing, X. Wang, Z. Zhang, Y. Chen, X. Huang, et al., Two-dimensional light-emitting materials: preparation, properties and applications, Chem. Soc. Rev. 47 (2018) 6128.
b) R. Dong, M. Pfeffermann, H. Liang, Z. Zheng, X. Zhu, J. Zhang, et al., Large-area, free-standing, two-dimensional supramolecular polymer single-layer sheets for highly efficient electrocatalytic hydrogen evolution, Angew. Chem., Int. Ed. 54 (2015) 12058–12063.
c) Z.-Q. Lin, J. Liang, P.-J. Sun, F. Liu, Y.-Y. Tay, M.-D. Yi, et al., Spirocyclic aromatic hydrocarbon-based organic nanosheets for eco-friendly aqueous processed thin-film non-volatile memory devices, Adv. Mater. 25 (2013) 3664–3669.
[18] a) K.S. Novoselov, A.K. Geim, S.V. Morozov, D. Jiang, Y. Zhang, S.V. Dubonos, et al., Electric field effect in atomically thin carbon films, Science 306 (2004) 666.
b) A.K. Geim, K.S. Novoselov, The rise of graphene, Nat. Mater. 6 (2007) 183.
c) Y.M. Lin, C. Dimitrakopoulos, K.A. Jenkins, D.B. Farmer, H.Y. Chiu, A. Grill, et al., 100-GHz transistors from wafer-scale epitaxial graphene, Science 327 (2010) 662.
d) F. Bonaccorso, Z. Sun, T. Hasan, A.C. Ferrari, Graphene photonics and optoelectronics, Nat. Photon. 4 (2010) 611.
e) Y.W. Son, M.L. Cohen, S.G. Louie, Half-metallic graphene nanoribbons, Nature 444 (2006) 347.
f) J. Maassen, W. Ji, H. Guo, Nano Lett. 11 (2011) 151.
[19] a) P. Kissel, R. Erni, W.B. Schweizer, M.D. Rossell, B.T. King, T. Bauer, et al., A two-dimensional polymer prepared by organic synthesis, Nat. Chem. 4 (2012) 287.
b) P. Payamyar, K. Kaja, C. Ruiz-Vargas, A. Stemmer, D.J. Murray, C.J. Johnson, et al., Synthesis of a covalent monolayer sheet by photochemical anthracene dimerization at the air/water interface and its mechanical characterization by AFM indentation, Adv. Mater. 26 (2014) 2052.
c) M.J. Kory, M. Wörle, T. Weber, P. Payamyar, S.W. van de Poll, J. Dshemuchadse, et al., Gram-scale synthesis of two-dimensional polymer crystals and their structure analysis by X-ray diffraction, Nat. Chem. 6 (2014) 779.
d) P. Kissel, D.J. Murray, W.J. Wulftange, V.J. Catalano, B.T. King, A nanoporous two-dimensional polymer by single-crystal-to-single-crystal photopolymerization, Nat. Chem. 6 (2014) 774.
e) T. Bauer, Z. Zheng, A. Renn, R. Enning, A. Stemmer, J. Sakamoto, et al., Synthesis of free-standing, monolayered organometallic sheets at the air/water, Interf. Angew. Chem. Int. Ed. 50 (2011) 7879.
f) Z. Zheng, L. Opilik, F. Schiffmann, W. Liu, G. Bergamini, P. Ceroni, et al., Synthesis of two-dimensional analogues of copolymers by site-to-site transmetalation of organometallic monolayer sheets, J. Am. Chem. Soc. 136 (2014) 6103.
[20] a) S. Kusaka, R. Sakamoto, Y. Kitagawa, M. Okumura, H. Nishihara, An extremely bright heteroleptic bis(dipyrrinato)zinc(II) complex, Chem. Asian J 7 (2012) 907–910.
b) M. Tsuchiya, et al., Asymmetric dinuclear bis(dipyrrinato)zinc(II) complexes: broad absorption and unidirectional quantitative exciton transmission, Chem. Commun 50 (2014) 5881–5883.
c) S.A. Baudron, Luminescent metal–organic frameworks based on dipyrromethene metal complexes and BODIPYs, Cryst. Eng. Comm. 18 (2016) 4671–4680.

[21] a) S. Skoulika, P. Dallas, M.G. Siskos, Y. Deligiannakis, A. Michaelides, Crystal structure and solid-state reactivity of a Cd(II) polymeric complex with acetylene-dicarboxylic acid, Chem. Mater. 15 (2003) 4576.
b) S. Schrettl, C. Stefaniu, C. Schwieger, G. Pasche, E. Oveisi, Y. Fontana, et al., Functional carbon nanosheets prepared from hexayne amphiphile monolayers at room temperature, Nat. Chem. 6 (2014) 468.
c) D.C. Milan, M. Krempe, A.K. Ismael, L.D. Movsisyan, M. Franz, et al., The single-molecule electrical conductance of a rotaxane-hexayne supramolecular assembly, Nanoscale 9 (2017) 355–361.
[22] R. Sakamoto, K. Hoshiko, Q. Liu, T. Yagi, T. Nagayama, S. Kusaka, et al., A photofunctional bottom-up bis(dipyrrinato)zinc(II) complex nanosheet, Nat Commun. 6 (2014) 6713.
[23] I.V. Sazanovich, C. Kirmaier, E. Hindin, L. Yu, D.F. Bocian, J.S. Lindsay, et al., Structural control of the excited-state dynamics of bis(dipyrrinato)zinc complexes: self-assembling chromophores for light harvesting architectures, J. Am. Chem. Soc 126 (2004) 2664–2665.
[24] R. Sakamoto, T. Iwashima, M. Tsuchiya, R. Toyoda, R. Matsuoka, J.F. Kögel, et al., New aspects in bis and tris(dipyrrinato)metal complexes: bright luminescence, self-assembled nanoarchitectures, and materials applications, J. Mater. Chem. A 3 (2015) 15357–15371.
[25] T. Bauer, Z. Zheng, A. Renn, R. Enning, A. Stemmer, J. Sakamoto, et al., Synthesis of free-standing, monolayered organometallic sheets at the air/water interface, Angew. Chem. Int. Ed. 50 (2011) 7879–7884.
[26] J.N. Coleman, M. Lotya, A. O'Neill, S.D. Bergin, P.J. King, U. Khan, et al., Two-Dimensional nanosheets produced by liquid exfoliation of layered materials, Science 331 (2011) 568.
[27] a) C.S. Diercks, O.M. Yaghi, The atom, the molecule, and the covalent organic framework, Science 355 (2017) 11585.
b) N. Huang, P. Wang, D. Jiang, Covalent organic frameworks: a materials platform for structural and functional designs, Nat. Rev. Mater 1 (2016) 16068.
c) J. Sakamoto, J. van Heijst, O. Lukin, A.D. Schlüter, Two-dimensional polymers: just a dream of synthetic chemists? Angew. Chem. Int. Ed 48 (2009) 1030–1069.
[28] A. Mal, R.K. Mishra, V.K. Praveen, M.A. Khayum, R. Banerjee, et al., Supramolecular reassembly of self-exfoliated ionic covalent organic nanosheets for label-free detection of double-stranded DNA, Angew. Chem. Int. Ed 57 (2018) 8443.
[29] L. Wang, H. Sahabudeen, T. Zhang, R. Dong, Liquid-interface-assisted synthesis of covalent-organic and metal-organic two-dimensional crystalline polymers, NPJ 2D Mater. Appl. (2018) 25.
[30] a) P. Payamyar, K. Kaja, C. Ruiz-Vargas, A. Stemmer, D.J. Murray, C.J. Johnson, et al., Synthesis of a covalent monolayer sheet by photochemical anthracene dimerization at the air/water interface and its mechanical characterization by AFM indentation, Adv. Mater 26 (2014) 2052–2058.
b) W. Dai, F. Shao, J. Szczerbiński, R. McCaffrey, R. Zenobi, Y. Jin, et al., Synthesis of a two-dimensional covalent organic monolayer through dynamic imine chemistry at the air/water interface, Angew. Chem. Int. Ed 55 (2016) 213–217.
[31] J. Liu, W. Zan, K. Li, Y. Yang, F. Bu, Y. Xu, Solution synthesis of semiconducting two-dimensional polymer via trimerization of carbonitrile, J. Am. Chem. Soc 139 (2017) 11666–11669.

CHAPTER 6

Interfaces in industrial applications

6.1 Introduction

In the previous chapters, I have analyzed the fundamental principles underlying the structure, properties, and nature of the interface and the synthesis and self-assembly of materials between two liquids. In this chapter, I will analyze and discuss the applications of nanomaterials, polymers, and polymer composites synthesized through liquid–liquid interfaces or upon application in two-phase systems in fields such as energy storage, solar energy conversion, light-harvesting, oil recovery/tracing, and sensors. Composite materials such as catalysts and self-assembled monolayers can be synthesized readily through this versatile methodology. Finally the chapter is presenting an extensive discussion on the role of liquid–liquid interfaces in niche applications such as enhanced oil recovery and their effect on catalytic reactions. These are applications that are becoming increasingly important with the discovery of numerous unconventional oil resources worldwide that promise to satisfy our energy needs for the next decades.

6.2 Energy applications: composite anode catalysts, self-assembled monolayers, and capacitive storage

In our energy-consuming society, alternative energy storage and energy-conversion devices are of fundamental importance due to constantly increasing energy needs and the dangers associated with some of the available energy sources, such as nuclear power and conventional/unconventional oil and gas resources. Fuel cells, solar cells, and supercapacitors are at the forefront of the current research promising to reshape the map of the storage and conversion of energy [1a]. Supercapacitors are electrochemical devices that store energy either through ion adsorption (the electrochemical double-layer capacitors) or through surface redox reactions (pseudocapacitors). They are considered as promising future replacements for batteries for energy storage and energy harvesting [1b].

Generation of Polymers and Nanomaterials at Liquid-Liquid Interfaces
DOI: https://doi.org/10.1016/B978-0-12-819491-1.00006-X

In fuel cells the intermetallic nanoparticles (NPs) that are based on noble metals, and transition metals have shown excellent catalytic activity toward methanol oxidation when they are used as electrodes [2]. The most important examples of such NP systems are the platinum-based NiPt, FePt, and $FePt_3$ intermetallic alloys [3]. These alloys have been demonstrated to be superior catalysts for methanol electrooxidation [4] and via incorporation of carbon particles, such as the Vulcan XC-72 [5]. Other carbon nanomaterials (e.g., nanotubes, graphene), as catalyst supports, are enhancing the efficiency of direct methanol fuel cells. These are hierarchical intermetallic alloys that have been used for methanol oxidation. However, despite these efforts, the rate of methanol electrooxidation is low due to the low utilization of Pt on conventional carbon support. This is related to the low electrochemically accessible surface area for the deposition of platinum particles. Conductive polymers have been proposed as candidates for a new type of support due to their unique properties: highly accessible surface area, high conductivity, air/thermal stability. Another class of nanomaterials that may increase the efficiency of these catalysts is fullerene "buckyballs." The latter are very efficient electron acceptors with a variable and rich range of redox potentials depending on the cage size and shape that may attach to the surface of noble metal NP through chemical functionalization [6].

Various liquid–liquid interfacial methods have been developed for the synthesis of conductive polymer composites for numerous energy-related applications. For example, Zhao and coworkers developed an in-situ interfacial polymerization method of pyrrole for deposition on carbon black particles. This was followed by codeposition of platinum and iron on the polypyrrole (PPy)–carbon composite to prepare a bimetallic platinum–iron/PPy–carbon-functional material that may be used as a catalyst [7]. These platinum–iron/PPy–carbon catalysts demonstrated improved catalytic activity for methanol oxidation compared with commercial and conventional catalysts comprising platinum and carbon composite particles. For the synthesis of composites, initially Vulcan XC-72 carbon black was treated with nitric acid at 100°C to enable decoration with hydroxyl and carboxyl groups on its surface. Subsequently the PPy–carbon support was synthesized by an in-situ interfacial method whereby the pyrrole monomer was dissolved in the organic phase and carbon particles were dispersed homogeneously in a mixture of isopropanol–water in the presence of iron nitrate as an oxidizer.

Besides PPy another example is the self-assembled monolayers of polyaniline (PANI) that have been synthesized through a two-phase experimental procedure reported by Sawall and coworkers. They demonstrated an interfacial polymerization method to form a two-dimensional (2D) array of PANI fibers grafted onto a gold substrate [8]. The first step involved the formation of a self-assembled monolayer from 4-aminithiolphenol molecules arranged on the gold surface through the well-known and very strong Au–S bond, which is typical for noble metals and thiols. In a second step, PANI nanofibers were synthesized directly onto the gold layer by placing the substrate at the liquid–liquid interface of a two-phase system comprising a dopant and aniline monomer. The substrate, treated previously with 4-aminothiolphenol, was placed at the interface with the amine groups heading toward the solution and arranged parallel, and aniline monomers diffused from the organic layer to the interface. Subsequently aniline monomers were protonated by the acidic aqueous layer to form an anilinium cation that attacked the aromatic ring and connected to the end of a growing polymer [9].

Polymeric nanofibers show high, easily accessible, and penetrable surface areas. They show advantageous behavior compared with their conventional counterparts and are expected to demonstrate improved storage of capacitive charge. More specifically a further increase in capacitive charge storage should be observed as their diameter decreases. For example, Manohar and coworkers reported a specific capacitance of 277 F/g. This value was observed for PANI–2-acryloamino-2-methyl-1-propanesulfonic acid (AMPSA) nanofibers of a mean diameter of 23 nm. Measurements were taken in aqueous 1.0 M camphor sulfonic acid electrolyte at a potential range of 0.4–0.5 V. To demonstrate comparative experiments, the authors used nonfibrillar PANI which was doped again with AMPSA and also obtained from the same reaction batch of PANI–AMPSA. Measurements gave a much lower capacitance of 11 F/g. These comparisons demonstrated the superior performance of fibrillar PANI compared with the bulk, randomly oriented polymer, with a much higher overall capacitance and greater symmetry in charge–discharge cycles. This remarkably high capacitance of PANI–AMPSA fibers may be related to other factors, for example, the fact that AMPSA is a reactive agent and undergoes free-radical polymerization in aqueous oxidative conditions, which are exactly the conditions for the synthesis of conductive polymers. In effect, under the conditions of this polymerization, a polymer from the doping agent may be formed as a byproduct and subsequently increases the available surface area [10a]. The capacitance and current density may be compared with other works in this field. Bora and Dolui dispersed graphene and an initiator in the aqueous phase and pyrrole in the oil phase to synthesize graphene−PPy composites for supercapacitors. They observed a specific capacitance of 260 F/g, the conductivity of 8.45 S/cm, and the current density of 100 mA/g [10b]. Thiophene−graphite oxide composites through a nitromethane−hexane system showed a specific capacitance of 99 F/g [10c]. The same authors developed PANI−nickel oxide composites synthesized again through an interfacial polymerization method [10d].

Interfacial polymerization of anilines has also been used for the synthesis of high-performance asymmetric supercapacitors, as reported by Shen and coworkers [11]. In their experimental work the aniline monomer was dissolved in dichloromethane, whereas the sulfonated graphene was dispersed in a water−isopropanol mixture. The interfacial self-assembly method led to excellent interfacial contact between PANI nanorods, sulfonated graphene nanolayers, and multiwalled carbon nanotubes (MWCNT), and enhanced cycling performance of asymmetric supercapacitors.

In general both small ions and surfactants have been used as counterbalancing ions in conductive polymers. However, Iekha and coworkers reported the use of polyoxometalates (POM) as doping agents in PANI. POMs, as counterbalancing ions, increase homogeneous charge separation on the macromolecule through the delocalization of their negative charges, thereby facilitating "electron hopping" and charge transfer between polymer chains [12−14]. An example of the influence of doping agents on the properties of conductive polymers is doping with POM ions in PPy, which has been utilized as a supercapacitor by White and coworkers [14]. Ma and coworkers synthesized PANI doped with POM and with silver NP. In general PANI has a very weak ultraviolet−visible (UV−vis) photo-response due to the strong and quick recombination of the photo-generated electron−hole pair. However, in that work, the polymer exhibited photo-responsive behavior toward UV radiation due to the efficient dissociation of the photo-induced hole. The electron transfer took place from the silver to the polymer and then to the POM, and the photo-current was measured for many cycles under UV irradiation of 365 nm [13b].

White and coworkers showed that the highest values of specific capacitance, energy, and power could be obtained with PPy electrodes incorporating polymer doped with 10-molybdo-2-vanadophosphoric acid. In addition to these findings PANI nanotubes and nanofibers synthesized through a soft template methodology also demonstrated very good electrochemical performance on lithium batteries and supercapacitors [15,16]. Another monomer used for the fabrication of pseudocapacitive electrodes through liquid–liquid interfaces was 2,5-dimethoxyaniline by Yuan and coworkers [16b]. They synthesized hierarchically ordered graphene oxide with PANI nanocomposites through an interfacial polymerization method. Similar effects have been observed for composite materials from hierarchically ordered graphene oxide and PANI [17] that exhibit high specific capacitance of 1,095 F/g at 1 A/g, an energy density of 24.3 W h/kg, and a maximum power density of 28.1 kW/kg. These graphene–PANI composites outperformed most other solid-state supercapacitors [18] thanks to the synergistic effects of the two components. The conductive polymer and graphene layers were arranged in an improved, hierarchical structure in the nanocomposite.

Before discussing the applications of supercritical fluids in underground environments, I will detail briefly their use for the templating synthesis of porous polymer materials. It is known that supercritical drying can be used for the generation of highly porous, fibrillar networks of biopolymers or conductive polymers (e.g., cellulose, PPy) with high surface area and good conductivity [19a]. Xu and coworkers employed 4-(2,3-dihydrothieno[3,4-*b*][1,4]dioxin-2-yl)-methoxybutane-1-sulfonate as a surfactant that would disperse 3,4-ethylenedioxythiophene (EDOT) to be polymerized later in an emulsion with $FeCl_3$ as the oxidizing agent. The aerogels exhibited relatively high conductivity (10^1 S/m), good electrochemical capacitance with cycling stability, and could absorb dyes and heavy metals [19b]. These types of materials can be used as electrodes in batteries; be employed in ion-exchange applications as exchange membranes with good mechanical stability and high porosity; use carbon dioxide (CO_2; an abundant greenhouse gas that scholars are striving to convert to useful chemicals). Other works on nanostructured polymer gels and conductive rod brushes, as well as hollow particles, reported their synthesis through liquid–liquid interfaces [20,21]. They have also been reported to be cost-effective materials for dye absorption, copper sensors, as well as having light emission and waveguide properties [22].

Another use of supercritical fluids that leads to ordered porous material is the synthesis of silicone oxycarbides (Si–O–C) for energy-storage applications. Among the various energy-related materials and devices, research has centered on lithium batteries due to modern technological needs. Graphite has been used as the anodic material, but alternative options are Si–O–C, which possesses a high specific capacity for the next generation of lithium-ion batteries [23]. However, their capacity fades quickly and they do not have sufficiently good rate performances, which hinder their application. Xia and coworkers used the microalgae *Nannochloropsis* as a "biological template" and carbon source for the synthesis of Si–O–C microspheres in the presence of supercritical carbon dioxide ($scCO_2$). Microalgae are abundant, renewable, and inexpensive. When the materials were tested as anodes in lithium-ion batteries, the microspheres exhibited a high reversible specific capacity of 450 mA h/g with a current density of 0.1 A/g over 200 cycles. Xia and coworkers developed a supercritical fluid-assisted biotemplating method for the synthesis of Si–O–C microspheres using microalgae as a biological template for advanced materials that can be used as anodes in lithium-ion batteries [23].

6.3 Use of surfactants, polymers, nanoparticles, and supercritical fluids in oil and gas resources

6.3.1 Enhanced oil recovery with supercritical fluids

The primary production of oil depends on the type of the reservoir, oil viscosity, and rock permeability, and averages only 30%–35%. However, for a typical gas field, the recovery is ≈80%. For decades engineers have been developing enhanced oil recovery methods for the exploitation of oil reservoirs that are limited in number and crude-oil capacity. Various enhanced oil-recovery methods have been developed: inert-gas injection, chemical flooding with surfactants, and thermal recovery using steam. The "wettability" of the underground rock formation plays a crucial part in the capillary pressure, permeability, and distribution of phases. Shohai-Kaveh and coworkers demonstrated these effects on CO_2–water–Bentheimer sandstone systems. They showed that, in contrast to a quartz [24a] system, the phase transition of CO_2 from a subcritical (2.83 MPa) to supercritical state (12.74 MPa) does not affect the wettability of Bentheimer sandstone [24b].

Liquid–liquid interfaces have been of great interest in the oil and gas industry, especially in enhanced oil-recovery processes using water or CO_2 emulsions [24c]. This involves the flooding of underground oil with liquids and solutions to reduce its viscosity and facilitate its removal from underground rock formations [25,26]. Enhanced oil recovery is very important in the upstream sectors of oil and gas industries. Consequently it is of great importance to understand how liquid–liquid interfacial systems between crude oil and various other fluids behave [27–30].

Water and carbon dioxide are the two most abundant solvents on Earth. They are nontoxic and nonflammable, and their interactions have been analyzed in Chapter 2, Nature of the Liquid–Liquid Interface. Due to gravity, segregation of emulsions of carbon dioxide with water or crude oil is not sufficiently stable. There have been many attempts to synthesize materials, surfactants, and functionalized NP that could thicken CO_2 and stabilize solutions [31]. In many instances the formation of stable emulsions is highly desirable, and many polymers and surfactants have been developed for $scCO_2$: polyvinyl ethyl ether, poly(1-decene), polyvinyl acetate [32], as well as block copolymers of fluoroacrylate and styrene [33]. In general fluorinated carbon atoms have been shown to have extraordinary solubility in $scCO_2$. This has been attributed to specific interactions between the polar C–F bond and quadrupole CO_2 [34]. However, fluorinated ligands are expensive and associated with numerous environmental concerns. To overcome these drawbacks that may hinder potential applications, many nonfluorinated $scCO_2$-philes have been developed. Most of them are based on acetate groups that have been shown to be excellent $scCO_2$-philes [35] due to an electron-withdrawing effect on nearby hydrogen atoms, which become more acidic and thus interact with CO_2 (which acts like a Lewis base [36]).

6.3.2 Role of asphaltenes in crude-oil emulsions

Hydrocarbons are the predominant organic molecules in crude oil, but aromatic molecules in underground resources must also interact with flooding fluids (water or CO_2). Asphaltene molecules are found in crude oil with a complex aromatic structure. To improve recovery efficiency, various molecular additives have been employed to reduce the interfacial tension of the oil–water mixture, either in oil extraction from the well or further separation at desalting plants. If the interfacial tension of the water–oil emulsion is very low, it will be very stable and the separation of water from crude oil in a subsequent step will be significantly more difficult. Usually this aqueous solution of underground water has very high salinity.

Besides the experimental methods, a plethora of computational methods, such as molecular mechanics, Monte Carlo, molecular dynamics

(MD), and dissipative particle dynamics (DPD), may provide insights on the behavior of water–oil mixtures. Crude oil is a mixture of thousands of different hydrocarbon molecules, so it is not a homogeneous one-phase system. The refinery products of oil can be classified as (1) paraffins, naphthenes, and aromatics; (2) paraffins, iso-paraffins, naphthenes, and aromatics; (3) paraffins, olefins, naphthenes, and aromatics; (4) paraffins, isoparaffins, olefins, naphthenes, and aromatics; or (5) saturates, aromatics, resins, and asphaltene. The latter seems to be the most common classification.

Asphaltenes are the most abundant components of a crude oil mixture. Asphaltenes consist of polycyclic aromatic rings, an aliphatic chain, and various heteroatoms (oxygen, sulfur, or nitrogen). The combination of these three subcomponents defines the properties of the crude oil mixture.

Asphaltenes and resins influence the stability of the emulsions significantly. Specifically asphaltenes are heavily used in petroleum processing since the oil–water emulsions are stabilized by asphaltenes with a decrease in the interfacial tension as the concentration of the asphaltenes increases. Hence the behavior of these molecules at the oil–water interface merits further investigation [37–39]. For these applications simulations indicate that the interface should be considered as the area where the density of water is 10%–90% of the density of its bulk phase. Asphaltenes are present in heavy oil and tend to form stable emulsions with water. These are invert emulsions, specifically water-in-oil emulsions stabilized (presumably) by these organic molecules. This interaction between asphaltenes with water has been investigated by Mikami and coworkers [40] (Fig. 6.1), who studied the interfacial behavior of asphaltenes at an oil–water liquid–liquid boundary through MD simulations. The behavior of asphaltenes in an emulsion appears to be complicated, and it has been shown that they can stabilize a water-in-organic-phase emulsion only if they are close to or above the point of incipient flocculation. An asphaltene film is 2D at low concentrations in the organic-phase (i.e., oil), but may form 3D arrays at high concentrations. Jian and coworkers [41], using the pendant drop method and MD simulations, quantified the effect of the concentration of asphaltene molecules on the interfacial tension of oil–water interfaces. Common organic solvents and a commercially available polyaromatic molecule, Violanthrone-79 (VO-79), were used as models of the oil phases, and the bulk and surface concentration were studied. These experiments revealed that it is the surface concentration,

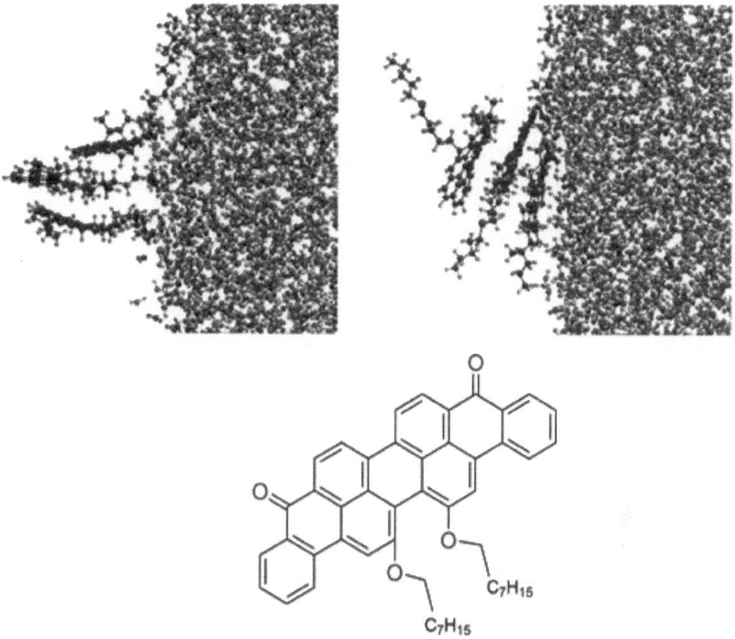

Figure 6.1 Configurations of asphaltene aggregates at the oil−water interface: (left) vertical structure and (right) parallel structure. *Reproduced with permission from Y. Mikami, Y. Liang, T. Matsuoka, E.S. Boek, Energy Fuels, 27, 1838 (2013). ©2013, American Chemical Society.*

and not the bulk concentration, that dictates the reduction of oil−water interfacial tension.

Due to their aromatic character, asphaltenes appear to be insoluble in heptane but soluble in aromatic solvents such as toluene [42]. Asphaltenes form aggregates in solution. These aggregates consist of ≈ 6 molecules, or bigger clusters of 8 molecules (as demonstrated by X-ray scattering, small-angle neutron scattering, centrifugation, and computational studies [43]) and precipitate from petroleum fluids as a highly viscous and sticky material [44−50].

Mikami and coworkers demonstrated that asphaltene molecules behave differently depending on the organic solvent employed. Asphaltenes migrate preferably to the oil phase if pure toluene is employed, whereas they tend to accumulate at the interface if pure heptane is used. This phenomenon may be correlated further with the interfacial tension of a two-phase system. If it contains a small number of asphaltene molecules, the interfacial tension is similar to that of a pure heptane−water interface,

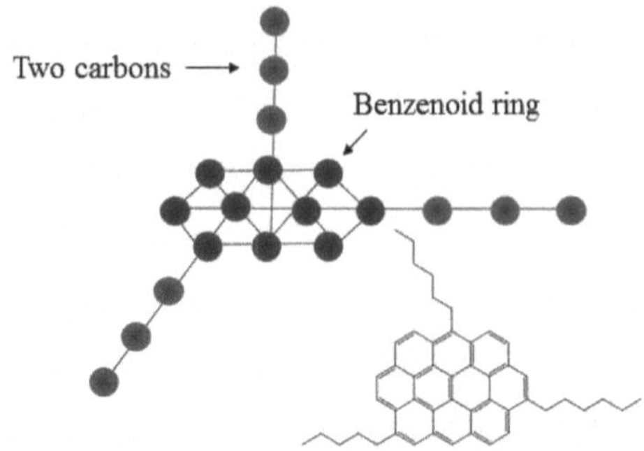

Figure 6.2 Molecular structure of an asphaltene molecule and the proposed coarse-grained bead imaging. The benzenoid ring in the ovalene aromatic ring comprises 10 fused aromatic hexagonal rings.

whereas the interfacial tension of a system containing a large number of asphaltene molecules is much reduced ($\approx$12 mN/m) because of the excellent distribution of asphaltene molecules that form a film at the oil–water interface (Fig. 6.1).

In Fig. 6.2, it is shown the coarse grain bead representation of an asphaltene molecule along with its molecular structure. The DPD-MD simulation contains four bead categories: Q (charged atoms), P (polar bonds), N (nonpolar bonds), and C (apolar bonds) with each group representing atoms or a liquid. Based on molecular orbital calculations the asphaltenes consist of 4–10 fused aromatic rings. In general the polycyclic aromatic hydrocarbons with 6–8 rings are the predominant types in petroleum asphaltenes.

Coarse-grained model molecules at the mesoscale level, by the use of DPD, have been used to study the preferred orientation of a model asphaltene molecule at the oil–water liquid–liquid interface, with toluene being used as a model of oil. Three orientations for asphaltene molecules were employed for the study of the interactions: horizontal to the oil – water interface; perpendicular to the oil – water interface; tilted 45° to the flat oil – water interface. This work demonstrated that the asphaltene molecule remains at the oil–water interface. The aromatic region of the asphaltene molecule is on the plane of the oil–water interface, whereas aliphatic chains are perpendicular to the oil – water interface and are headed (as expected) toward the organic phase, with the same orientation for higher surface coverage. Steric interactions play a significant part, with asphaltenes heading to the bulk phase and some remaining at the oil – aqueous interface, interacting through pi – pi stacking interaction of the aromatic rings.

Various organic molecules have been used as surrogate samples for asphaltenes, with the VO-79 to be the most common [51]. Nanoparticles such as pyroxene ($NaFeSi_2O_6$) also known as aegirine have also been used for the adsorptive removal of VO-79 from emulsions.

An overview of the use of interfacial systems for recovery of "unconventional oils" was published by He and coworkers [52a]. They discussed the role of surfactants, acids and bases, and salts (e.g., NaCl) on oil–water mixing with the possibility of enhanced oil recovery. Further analyses on the influence of temperature, pressure as well as the addition of nanofluids on the contact angle between rock and oil and miscibility between water and oil were presented by the authors. The term "unconventional oil" denotes heavy oil resources with very high density and viscosity. Bitumen has a density of 1050 kg/m^3 and is highly viscous ($> 10^3$ Pa s). The amount of unconventional resources is huge (1.7 trillion barrels of bitumen as tar sands in Canada). The properties of crude oil worldwide vary significantly, with Canadian Oil Sands reaching a viscosity of 10^6 Pa s.

6.3.3 Cosolvents in lignin removal and cellulose treatment

Biomass that contains lignin and cellulose is an alternative source of energy compared with oil and gas. Treatment with various solvents is an efficient way to separate the components of this biomass. Mixtures of cosolvents have been used for optimal separation. Examples include alcohols, ketones, ethers, and ionic liquids. Mostofian and coworkers employed a mixture of tetrahydrofuran–water cosolvent to produce a high amount of lignin since tetrahydrofuran and water phase separate at the surface of cellulose fibers. They demonstrated that the addition of tetrahydrofuran significantly increased the deconstruction of cellulose [52b]. It is also found that the addition of a strong acid, such as the sulfuric acid, in dilute concentrations can be a pretreatment catalyst beneficial for the lignin extractions since it reduces the activation energy of the hydrolytic cleavage of glycosidic bonds. On the other hand a high acid concentration might be detrimental due to the decomposition of the sugars and the lignin. In general experimental work including electron microscopy experiments in conjunction with extensive simulations has shown that the delignification and the chemical conversion occur and lead to solubilization of lignin and cellulose, but the exact mechanism of action is controversial and not yet fully elucidated.

6.4 Catalysts that operate in liquid–liquid interfaces and water-in-oil emulsions

One of the major drawbacks of one-phase, single-solvent reactions, and two-phase systems, especially on the industrial scale, is inhomogeneous mixing with heat-transfer limitations and insufficient interaction between reagents. Surfactants can be employed to overcome these obstacles, but they increase the cost of the whole procedure on the industrial scale. The catalytic effect of micelles was demonstrated a long time ago. The first examples were supramolecular aggregates with micelles behaving similarly to enzymes because they can trap and isolate species from their environment. Many nanomaterials, including chalcogenides, have been employed as catalysts and demonstrated a catalytic activity through the formation of emulsions. For example, MoS_2 is a semiconductor that is synthesized through sonolysis of hexacarbonyl molybdenum and yellow sulfur in hexadecane. The emulsions of MoS_2 with a nonionic surfactant, the Tergitol NP-8 in polyethylene glycol (PEG) or oil, has been demonstrated to lead to the catalytic synthesis of alcohols with very good yields [53].

Interfaces have also been employed in phase-transfer catalytic reactions, whereby the migration of reagents is facilitated through the expansion of the interfacial area [54a]. They are divided into two main categories: normal-phase transfer catalysts (with the use of quaternary ammonium and phosphonium surfactants) and inverse-phase transfer catalysts with cyclodextrin (CD) and calixarenes. A newly developed, tailor-made surfactant, polyethyleneglycol ubiquinol sebacate (Fig. 6.3), has also been used as a platform for catalysis. The catalytic reaction is rhodium-catalyzed

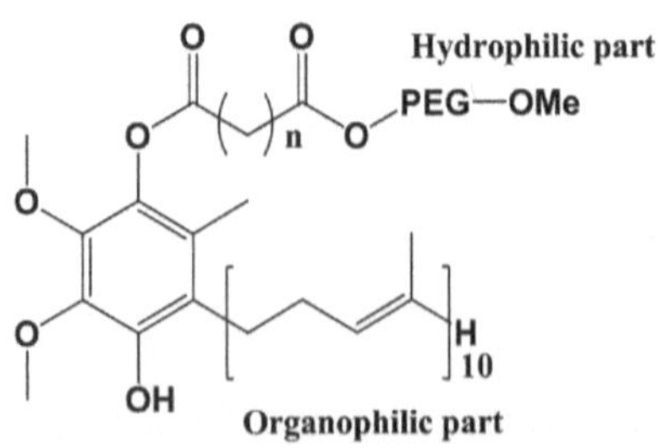

Figure 6.3 Chemical structure of a tailor-made and suitably modified multifunctional surfactant. Polyethylene glycol ubiquinol sebacate was used as the model.

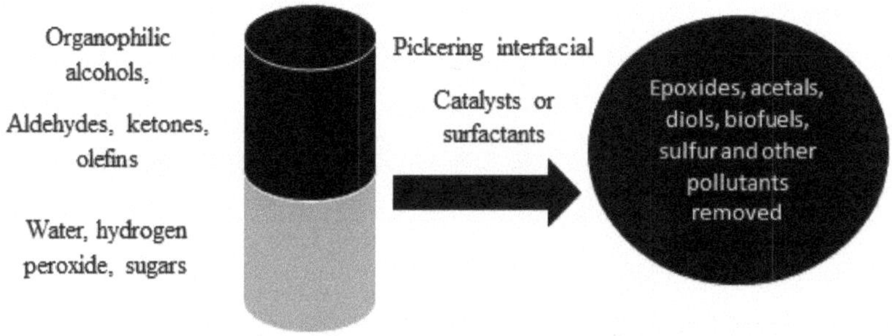

Figure 6.4 The different catalytic systems that are derived from or take place at liquid–liquid interfaces. The major products that are obtained from these processes are also indicated.

asymmetric conjugate addition [54c]. Fig. 6.4 summarizes the different catalytic systems from liquid–liquid interfaces.

An overview of the particles and liquid–liquid interface systems that have been used in catalysis are listed in Table 6.1.

This can be of particular importance if the products are unstable under specific conditions and a range of byproducts is produced, which can be separated readily through two phases. NP stabilize emulsions and, if catalytically active particles are incorporated, the liquid–liquid interface can be a medium for catalysis, as demonstrated by Crossley and coworkers [55]. They incorporated palladium NP (which are catalytically active) on a silica–CNT hybrid (Fig. 6.5). Then they tested their activity for biomass refining chemistry: hydrodeoxygenation of a phenolic compound and hydrogenation and esterification of an aldehyde. These reactions occur in the aqueous phase. Palladium particles, with their electron-rich surface sites, have also been used for biofuel upgrading among other catalytic reactions. Their wettability [57] as well as their surface chemistry and functionalization [58] affect their activity. Yang and coworkers as well as Xu and coworkers developed ruthenium and palladium catalysts for biofuel upgrading and hydrodeoxygenation of the bio-oil vanillin (4-hydroxy-3-methoxybenzaldehyde) [59]. They tested vanillin, a component of oil pyrolysis obtained through the fragmentation of lignin. This compound exhibits partial solubility in the organic and aqueous phase. Using this method, the catalytic conversion was achieved in the aqueous and organic phase. Reaction products migrated to the two phases. Ruthenium catalysts were supported by CNT, and the reaction conditions were relatively mild (1 MPa, 150°C). When the pressure was changed,

Table 6.1 Catalytic systems through liquid–liquid interfaces.

Type of system	Catalysis	Type of catalysis	Components	Phases
Demixed system	Phase-transfer catalysts	Normal phase-transfer catalysis	Opium salts, crown ethers	One or two phases
		Inverse phase-transfer catalysis	Cyclodextrins, calixarenes	One or two phases
		Phase-boundary catalysis	Amphiphilic zeolites	Liquid–liquid interfaces
Nanomaterials (1–100 nm)	Micellar catalysis	Micellar-assisted and phase-transfer catalysis	Surfactants	Liquid–liquid interfaces or both phases
	Microemulsion catalysis	Microemulsion and inverse-transfer catalysis	Crown ethers	
Microparticles (0.1–100 μm)	Surfactant emulsion catalysis	Surfactant-, interfacial- or surfactant-assisted catalysis	Surfactants, polymers, liquid–liquid interfaces	Liquid–liquid interfaces or both phases
	Pickering emulsion catalysis	Pickering-assisted catalysis	Catalytic particles, amphiphilic particles	

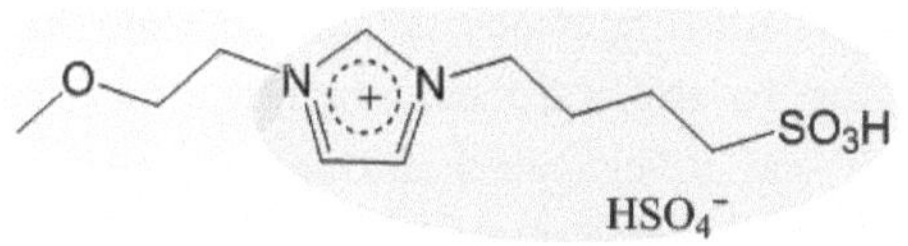

Figure 6.5 Pickering emulsion interfacial catalysts and a thermosensitive acidic imidazolium salt used for biodiesel production. *Adapted from M. Pera-Titus, L. Leclercq, J.-M. Clacens, F. De Campo, V. Nardello-Rataj, Angew. Chem. Int. Ed., 54, 2006 (2015) [56]. Copyright: Wiley-VCH.*

so did the catalytic activity. At 0.4 MPa, the conversion for the hydrodeoxygenation of vanillin was 50%, at 0.6 MPa it was 66%, and at 0.8 MPa it was 82%. Other works on the catalytic treatment of biofuel have involved reactions undertaken in a NiMo and CoMo intermetallics, which serve as Pickering emulsifiers (as indicated by fluorescence labeling agents). Pickering emulsions lead to the higher interfacial area and increase the concentration of hydrogen at the interface because hydrogen has a higher solubility in organic solvents than in water. Fluorescence imaging has shown catalysts to be located at the surface of droplets. The products were *m*-cresol, *p*-methoxyphenol, or vanillin alcohol at 50°C, with a remarkable reaction yield of 98%. While *m*-cresol and *p*-methoxyphenol migrate to the organic phase, vanillin alcohol migrates to the aqueous phase [59b]. In the work by Xu and coworkers, using 3-methyl-1-butylpyridine dicyanamide in combination with silica as a template led to 100% conversion of vanillin and 100% selectivity for 2-methoxy-4-methylphenol [59a]. Optical microscope images of the water-in-oil emulsion formed by a mixture of decalin and water in the presence of 5 wt.% palladium single-walled CNT silica nanohybrids and transmission electron micrographs of nanohybrids of palladium show noble metal particles to be located preferentially over the silica surface rather than CNT as shown by Crossley and coworkers [55].

Furthermore the Pickering emulsions have been extensively used for the petroleum industry where a green approach to control emulsion inversion in basic conditions is highly desirable anticipated. Liu et al. reported a phase inversion of a Pickering emulsion through the controlled wettability of silica NPs by *N*-(2-((2-aminoethyl)amino)ethyl)octadecenamide (denoted C18PDA) after the addition of Na_2CO_3 salt [60]. The authors observed a transitional phase inversion from an oil-in-water (O/W) emulsion to a water-in-oil (W/O) emulsion after increasing the addition of the C18PDA surfactant. The hydrophobicity of the silica NPs is reduced by salinity and its property can be used for creating suitable O/W emulsions due to the reduced interactions between the NPs and the C18PDA. This was investigated through studies of turbidity, surface tension, contact angle, Fourier transform infrared spectroscopy with thermogravimetric analysis, zeta potentials, etc.

Similar composite systems have been employed as Pickering interfacial catalysts for biphasic reactions [60], with the composite materials combining hydrophobic carbon structures with hydrophilic particles. Other examples include the MWCNT grown on alumina and silica. They are used for hydrogenation and partial-oxidation reactions [61]. The gold-based amphiphilic carbon catalysts that have found applications for biofuel upgrading include rubidium composites with nanotubes, as well as iron oxide and titania NP combined with nanotubes [62]. Copper particles have also been tested as catalysts and have been compared with palladium as catalyst supports on silica, alumina, and CNT substrates. For a detailed analysis of the different two-phase systems used for catalysis, see the review by Pera-Titus and coworkers.

Another interesting report is by Fang and coworkers on the amphiphilic Aquivion—carbon composites that can be used as interfacial acid catalysts for the acetalization reaction of C_{12}-aldehyde (Fig. 6.4) [63]. Aquivion is a newly developed category of perfluorosulfonic superacid resins with sulfonyl fluoride vinyl ether side chains, and the carbon catalysts are synthesized through a one-pot hydrothermal carbonization of cellulose gum [64]. Pera-Titus and coworkers focused on interfacial catalysts using imidazolium salts (Fig. 6.4).

Examples of the use of rare earth metal catalysts include cerium catalysts for the transesterification of soybean oil. Other rare earth metals have been used, and activity is reduced in the order Ce > Sm > Gd > La due to higher catalyst deactivation, and micelles may be used to promote the reaction [65]. Other rare earth and transition metal interfacial catalysts include copper and cerium dodecyl sulfates, which are Lewis acid surfactant catalysts [66]. Finally the rhodium-catalyzed hydroformylation of higher alkenes in Pickering emulsions based on a mixture of native α-CD and high molecular weight PEG have been reported [67]. Another surfactant-combined catalytic system is an ionic liquid consisting of a POM anion and a triblock copolymer (P123)-functionalized imidazolium cation [68], which is an efficient catalyst in the esterification of lauryl alcohol with acetic acid thanks to the formation of an emulsion [69].

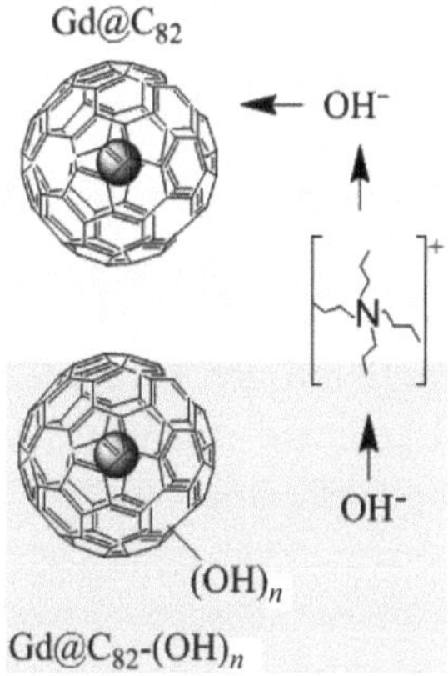

Figure 6.6 Phase-transfer catalysis for the synthesis of water-soluble gadolinium endohedral metallofullerenes (Gd@C_{82}). Particles are decorated with hydroxyl groups, and pristine fullerenes are soluble in toluene (upper phase). Subsequently they moved to the aqueous phase. *Adapted from M. Mikawa, H. Kato, M. Okumura, M. Narazaki, Y. Kanazawa, N. Miwa, et al., Bioconj. Chem. 12 (2001). Copyright: American Chemical Society.*

Phase-transfer catalysts have been used extensively. A particularly interesting example is the use of quaternary ammonium phase-transfer catalysts for the synthesis of water-soluble gadolinium endohedral metallofullerenes, (structure: Gd@C_{82}) (Fig. 6.6) [70a]. The gadolinium encapsulating fullerenes is synthesized through arc discharge evaporation of metal oxide-doped composite graphite. This interfacial functionalization method was developed by the Shinohara research team at Nagoya University, and water-soluble fullerenes can be very important for magnetic resonance imaging (MRI) applications [70b]. They also tested these particles as T1-weighted MRI agents in the lungs, liver, spleen, and kidneys of CDF1 mice in vivo.

Polyoxometalates are anionic clusters soluble in water but are less compatible with organic solvents. To be compatible with liquid–liquid interfaces, amphiphilic POM has been synthesized following two distinctive pathways, one of which is when POM anions are bound to a cationic surfactant. Li and coworkers reported the synthesis of a POM-based catalyst, a $[p\text{-}C_5H_5N(CH_2)_{15}CH_3]_3[PW_4O_{16}]$ system, that can carry out cyclohexene oxidation in the presence of hydrogen peroxide using water–chloroform interfacial system [68]. For the activity of the complex, the catalyst had to be broken down into smaller species such as $\{(PO_4)[WO(O_2)_2]_4\}_3$, $\{(PO_4)[WO(O_2)_2(H_2O)]\}_3$, and $\{[PO_3(OH)][WO(O_2)_2]_2\}_2$, all of which have good affinity with chloroform. Subsequently the oxygen from the tungsten compound $[W_2O_2(O_2)_4]$ reacts with the olefin to form a peroxide. Other POM emulsifiers can have the structure $Q_3[PW_{12}O_{40}]$ – where "Q" is quaternary ammonium cations. Applications of systems such as $[(C_{18}H_{37})_2N\text{-}(CH_3)_2]_3[PW_{12}O_{40}]$ could be the oxidation of sulfur-containing molecules to sulfones in diesel fuels, and the oxidation of alcohols to ketones using hydrogen peroxide, as demonstrated by Ding and coworkers in an ethyl acetatate–olefin–H_2O_2 system [68b].

The advantage of using surfactant-stabilized systems is the increase in the interfacial area between the two-phases at the nanoscale and microscale. However, in many cases the recovery of products and catalysts is quite rigorous and difficult. Nanomaterials, colloidal particles, polymers, and microgels can stabilize emulsions in the presence of water and oil, but they lead to three-phase liquid–solid–liquid media. CD may be used as emulsifiers instead of long-chain surfactants for catalytic reactions (Fig. 6.7) [71]. The presence of oil molecules (e.g., soybean oil), various *n*-alkanes and *n*-alkanols leads to the nonpreferred formation of insoluble oil–CD inclusion complexes [72]. Nardello-Rataj and coworkers presented various catalytic applications of this type of emulsion. These applications included the oxidation of alkenes, alcohols, and organosulfur compounds [73] with the water-soluble POM catalysts $Na_3(PW_{12}O_{40})$ and H_2O_2 to be used as the oxidant for the conversion of cyclooctene to the corresponding epoxide. This catalytic conversion was found to be nearly two orders of magnitude higher than the corresponding conversion in the absence of a surfactant.

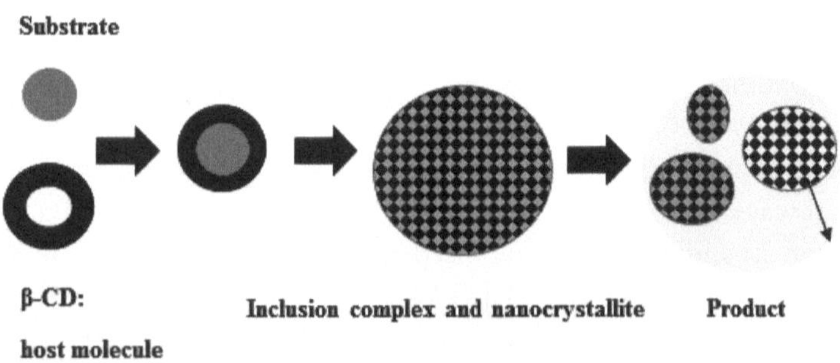

Figure 6.7 Interfacial catalytic activity in a Pickering emulsion and two-phase system with β-CD and nanocatalysts. *Adapted from M. Pera-Titus, L. Leclercq, J.-M. Clacens, F. De Campo, V. Nardello-Rataj, Angew. Chem. Int. Ed. 54, 2006 (2015) [56].*

I will elaborate on the applications of interfaces in the food science, starting with the well known in the fitness industry whey which possesses gelling properties. In whey the major protein is a-lactoglobulin, a globular protein which contains 162 amino acids and has a molecular weight of approximately 18,300 Da. Various works have been dedicated on the interaction between protein-adsorbed molecules at the air–water and oil–water interface. Interfacial rheology and optical methods have been applied to bubbles or drops and the gelation of whey protein globules took place at the oil–water interface of an aqueous solution of protein in the work of Rodrıguez Patino et al. [74]. As indicated in Fig. 6.8, the gelation takes place consecutively at 40°C, 60°C, and 80°C.

It is important to notice here that at room temperature a-lactoglobulin exists as a dimer in water at a pH between 5.5 and 7.5 with intermolecular disulfide bonds. On the other hand after heating a-lactoglobulin solutions, several intramolecular and intermolecular changes lead to the dissociation of the dimer into its monomer units, and at temperatures above 60°C the monomers undergo a denaturation step. The authors aimed to introduce surface dilational measurements as a technique to investigate the viscoelastic properties of heat-set induced whey protein at the oil–water interface and to study the heat-induced interfacial aggregation.

We have already discussed the rules that govern the electron and mass transfer across the liquid–liquid interfaces. The electrochemistry at the interface presents some advantages in electroanalysis. The molecular and ionic species are susceptible to transfer across the interface and has been applied to silver cations [75], dopamine [75b], Cr(VI) [75c], aminoacids [75d], peptides [75e], proteins [75f], carbohydrates [75g], and dendrimers [75i]. It can also be used as potential controlled selective extraction of ions by tuning the applied potential difference among the two immiscible liquids. In the work by Herzog et al. [75j], aspartame and acesulfame K were the two food additives that were tested at the electrochemical interface at a custom made cell [75k]. A typical organic electrolyte salt can be prepared by metathesis of bis-(triphenylphosphoranylidene) ammonium chloride (denoted as BTPPA + Cl−) and potassium tetrakis (4-chlorophenyl)borate (K + TPBCl−) to obtain the BTPPATPBCl.

The increase in the interfacial area and catalytic effect of micelles may be expanded by using immobilized enzymes in Pickering or micellar emulsions. Enzymes are the selective catalysts of biological systems. They have a key role in reactions taking place in living organisms through unparalleled selectivity while carrying out their actions in aqueous environments.

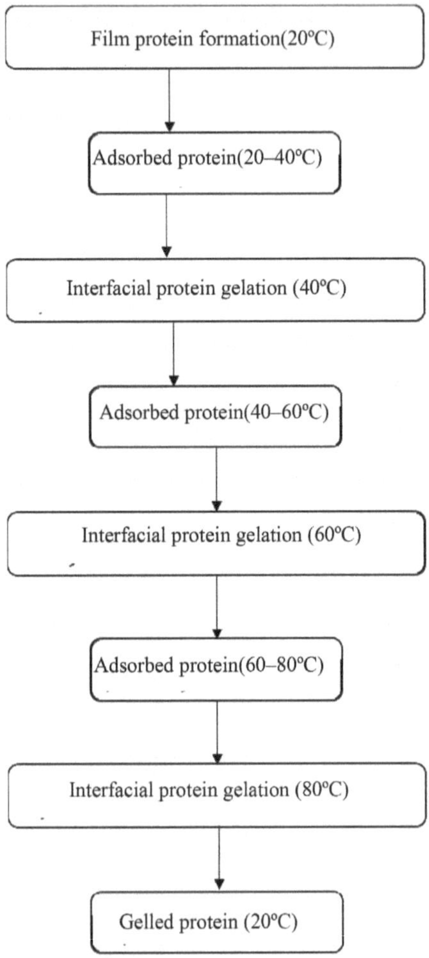

Figure 6.8 A schematic representation for the gelation procedure of the whey protein isolate. Based on the work by Rodrıguez Patino et al. To analyze the protein gelation at the oil−water interface, the protein solution is placed in a syringe and the oil in a cuvette. The viscoelastic characteristics of the protein film were monitored over time. *J.M. Rodrıguez Patino, M. Rosario Rodriguez Nino, C.C. Sanchez. Dynamic interfacial rheology as a tool for the characterization of whey protein isolates gelation at the oil-water interface. J. Agric. Food Chem. 47, 3640 − 3648 (1999).*

Incorporation of enzymes within emulsion droplets from colloidal NP or polymer-stabilized systems may have an impact on enzymatic activity [76,77]. The enzymes may be located at the water−oil interface, with or without surfactants or stabilizing NP, but there may be some drawbacks in that case. Various strategies have been proposed to overcome the loss of enzymes from droplets. For example, lipase pancreatic enzymes (which catalyze the conversion of fats to fatty acids) have been attached to mesoporous silica after the silica particles were rendered hydrophobic though functionalization [78]. Similar strategies have been employed for the esterification reactions or hydrolysis of methyl esters with very high conversion rates ($\leq$90%) [79]. Another example is the incorporation of lipase enzymes in polyamidoamine dendrimers for the hydrolysis of olive oil [80]. There are various reports on dendrimer-functionalized colloidal catalysts [81]. Kim and coworkers synthesized multifunctional, multicatalytic composite colloids through a layer-by-layer approach. Recyclable colloids with various catalytic properties were formed from a ligand-exchange process between a metal or metal oxide NP functionalized with hydrophobic group and amine-terminated dendrimers. Colloidal particles were dispersed readily in various organic solvents such as chloroform, toluene, hexane, and triethylamine, whereas an NH_2 dendrimer-functionalized SiO_2 colloid could be dispersed in polar media such as water and ethanol [81].

These multifunctional systems, synthesized through a layer-by-layer approach, exhibit high catalytic activity as well as good dispersion and recycling stability in organic and aqueous environments. A wealth of different metal oxide NP with the desired catalytic performance can become part of multilayered blocks with dense packing. They may have many (and possibly controllable) catalytic properties. Also by varying the thin external layer, different solubilities may be achieved. Examples of these core-shell colloids include palladium, iron oxide, and titanium dioxide catalysts. They have excellent performance for the oxidation of various organic molecules such as 3,3′,5,5′-tetramethylbenzidine, as photocatalysis in aqueous solutions, and even for complex organic reactions (e.g., the Sonogashira coupling reaction can elicit a remarkable 99% yield in organic solutions). This reaction involves the formation of carbon—carbon bonds between aryl (or vinyl) halide and terminal alkynes with palladium or copper catalysts. This reaction has been used for the synthesis of precursors for pharmaceuticals and natural products, and various organic materials [82] such as the anticancer drug ellipticine from ethyl 2-(4-pyridyl)acetate through the Sonogashira coupling reaction [83]. Palladium NP coated with amine- and NH_2-based dendrimers and tetraoctylammonium bromide with multilayered structures can be well dispersed in toluene, hexane, chloroform, or triethylamine [84].

Zhu and coworkers studied different POM species for the oxidation of cyclohexene, cyclohexanol, cyclohexanone, and 1,2-cyclohexanediol to adipic acid without organic solvents or phase-transfer catalysts [85a]. They proposed a catalytic cycle for biphase oxidation in an emulsion. The dimeric species $Q_2M_2O_3(O_2)_4$ – where "Q" is quaternary – is converted to the monomer $Q[M(O_2)]$ with H^+ in the presence of excess H_2O_2 [85b]. The long aliphatic chain inherits molecules with a lipophilic nature, which facilitates the transfer of the active oxygen between aqueous and organic media, with chains trapping peroxotungstate and peroxomolybdate anions carrying active oxygen into the organic phase. A detailed analysis of a catalytic reaction employing POM for the oxidation of cyclohexene, cyclohexanol, cyclohexanone, and 1,2-cyclohexanediol toward adipic acid has been shown by Zhu and coworkers [85a].

For industrial applications, continuous-flow two-phase systems may be advantageous for development due to their increased efficiency and productivity, safety, and ease of control and manipulation. Binks and coworkers [86] developed continuous-flow biphase mixtures with oil-phase-containing reagent molecules and a hydrophilic the catalyst. They demonstrated, in a column reactor, that their system is efficient in sulfuric acid-catalyzed addition reactions, a ring-opening reaction, and an enzyme-catalyzed chiral reaction. A red dye, Sudan I, which is soluble in octane, is used to visualize the flow of the system. Sulfuric acid could be replaced by heteropolyacids (e.g., $H_3PW_{12}O_{40}$), an amine (tetraethylene-pentamine), a trisodium salt of tri-(m-sulfonphenyl) phosphine, or the enzyme *Candida antarctica* lipase B. The authors developed an equation that can be used to clarify the relationship among the flow rate of the oil phase and the droplet size at the interface. They also studied the ring opening of an epoxide, and achieved the conversion of 87%–99%.

Liquid–liquid interfaces have been also used in electroanalytic tools, ion selective electrodes, and chemical sensors. Electrochemical architectures have been used for the detection of the β-blocker propranolol, while Morita and coworkers provided a detailed mechanistic view of the microscopic barrier mechanism of ion transport across liquid – liquid interfaces through molecular dynamics simulation studies. In other reports insulin was directly detected at such microinterfaces by adsorptive stripping voltammetry [87].

6.5 Outlook and perspectives

The diffusion of charge and mass as well as phase transfer through the interface of immiscible liquids has been utilized extensively and exploited to develop novel methodologies for the self-assembly of NP at 2D-restricted space and generation of conductive polymer nanostructures. Numerous reports have detailed the use of the interfacial polymerization of aniline, pyrrole, and EDOT. This has been demonstrated to be a powerful tool for the synthesis of ultra-small and highly crystalline nanostructures, including nanofibers and nanoneedles, with superior sensor properties. Other applications that can be envisioned with these materials are nanoelectronics and energy-storage materials. Furthermore liquid–liquid interfacial reactions can be seen as a versatile method for the synthesis of composite materials [88]. These are the technological applications stemming from the fascinating properties of conductive polymers. Interfacial polymerization provides invaluable insights in the study of the fundamental chemical and physical aspects underlying mass and electron transfer through these "imaginary membranes" and the evolution of nanostructures and self-assembled arrays at the interface.

Interfacial polymerization presents several advantages and versatilities compared with conventional one-phase polymerization: (1) the slow reaction rate leads to more crystalline and ordered polymer NP assemblies; (2) the ability to employ organophilic and hydrophilic surfactants and monomers; (3) formation of thin films with tunable thickness; and (4) the synthesis of hierarchically ordered composite materials with CNT, graphene, or silver NP. PANI, PPy, and poly(3,4-ethylenedioxythiophene) are the most commonly applied conductive polymers. Besides commonly used monomers, many other functional monomers, metal-organic framework precursors, and NP can be polymerized or form functional oligomers, nanocomposites or thin films with oxidative or polycondensation methods [89] at an interface. This strategy can lead to unique nanostructures [90,91] or films with new electrochemical, electrochromic, photoluminescence or photo-conductive, redox or electronic transport properties and materials suitable for interfacial transport [92]. This approach can open new perspectives and pathways in biomedical engineering [93,94], polymer chemistry, and physics [95,96]. In the oil and gas industries, this approach can aid improved and enhanced extraction of oil from wells that would otherwise consume a significant amount of time and resources.

References

1. a) J. Andrews and N. Jelley in Energy Science, 2nd Edition, Oxford University Press, Oxford, UK.

b) Y. Yin, K. Feng, C. Liu, S. Fan, A polymer supercapacitor capable of self-charging under light illumination, Journal of Physical Chemistry C 119 (2015) 8488.
c) P. Simon, Y. Gogotsi, Materials for electrochemical capacitors, Nature 7 (2008) 845.
2. B.M. Leonard, Q. Zhou, D. Wu, F.J. Di Salvo, Facile synthesis of ptni intermetallic nanoparticles: influence of reducing agent and precursors on electrocatalytic activity, Chemistry of Materials 23 (2011) 1136.
3. C. Wang, D. van der Vliet, K.L. More, N.J. Zaluzec, S. Peng, S. Sun, H. Daimon, G. Wang, J. Greeley, J. Pearson, A.P. Paulikas, G. Karapetrov, D. Strmcnik, N.M. Markovic, V.R. Stamenkovitz, Multimetallic Au/FePt3 nanoparticles as highly durable electrocatalyst, Nano Letters 11 (2011) 919.
4. C. Xu, Q. Li, Y. Liu, J. Wang, H. Geng, Hierarchical nanoporous ptfe alloy with multimodal size distributions and its catalytic performance toward methanol electro-oxidation, Langmuir 28 (2012) 1886.
5. G. Wu, L. Li, J.H. Li, B.Q. Xu, Polyaniline-carbon composite films as supports of Pt and PtRu particles for methanol electrooxidation, Carbon 43 (2005) 2579.
6. Z. Bai, L. Yang, Y. Guo, Z. Zheng, C. Hu, P. Xu, High-efficiency palladium catalysts supported on ppy-modified C60 for formic acid oxidation, Chemical Communications 47 (2011) 1752.
7. H. Zhao, L. Li, J. Yang, Y. Zhang, H. Li, Synthesis and characterization of bimetallic Pt−Fe/polypyrrole−carbon catalyst as DMFC anode catalyst, Electrochemical Communications 10 (2008) 876.
8. a) D.D. Sawall, R.M. Villahermosa, R.A. Lipeles, A. Hopking, R. Hopkins, Interfacial polymerization of polyaniline nanofibers grafted to au surfaces, Chemistry of Materials 16 (2004) 1606.
b) N.U. Day, M.G. Walter, C.C. Wamser, preparations and electrochemical characterizations of conductive porphyrin polymers, Journal of Physical Chemistry C 119 (2015) 17378.
9. X. Zhang, R.C.Y. King, A. Jose, S.K. Manohar, Nanofibers of polyaniline synthesized by interfacial polymerization, Synthetic Metals 145 (2004) 23.
10. a) A.V. Caramyshev, E.G. Evtushenko, V.F. Ivanov, A.R. Barceló, M.G. Roig, V.L. Shnyrov, R.B. van Huystee, I.N. Kurochkin, A.K. Vorobiev, I.Y. Sakharov, Synthesis of conducting polyelectrolyte complexes of polyaniline and poly(2-acrylamido-3-methyl-1-propanesulfonic acid) catalyzed by pH-stable palm tree peroxidase, Biomacromolecules 6 (2005) 1360.
b) C. Bora, S.K. Dolui, Interfacial synthesis of polypyrrole/graphene composites and investigation of their optical, electrical and electrochemical properties, Polymer International 63 (2013) 1439.
c) C. Bora, R. Pegu, B.J. Saikia, S.K. Dolui, Polymer International, Synthesis of polythiophene/graphene oxide composites by interfacial polymerization and evaluation of their electrical and electrochemical properties 63 (2014) 2061.
d) C. Bora, A. Kalita, D. Das, S.K. Dolui, P.K. Mukhopadhyay, Preparation of polyaniline/nickel oxide nanocomposites by liquid/liquid interfacial polymerization and evaluation of their electrical, electrochemical and magnetic properties, Polymer International 63 (2013) 445.
11. J. Shen, C. Yang, X. Li, G. Wang, High-performance asymmetric supercapacitor based on nanoarchitectured polyaniline/graphene/carbon nanotube and activated graphene electrodes, ACS Applied Materials Interfaces 5 (2013) 8467.
12. C.P. Lekha, S. Subramanian, D.P. Padiyan, Investigation of pseudocapacitance effect and frequency dependence of ac impedance in polyaniline−polyoxometalate hybrids, Journal of Materials Science 44 (2009) 6040.

13. a) A. Manivel, A.M. Asiri, K.A. Alamri, T.L. Villarreal, S. Anandan, Interfacially synthesized PAni–PMo12 hybrid material for supercapacitor applications, Bulletin Materials Science 37 (2014) 861.
    b) H. Ma, S. Yan, X. Pu, X. Shao, Y. Li, J. Gong, Y. Deng, Freezing-mediated polymerization of Ag nanoparticle-embedded polyaniline belts with polyoxometalate as doping acid exhibiting UV-photosensitivity, RSC Advances 6 (2016) 46475.
14. A.M. White, R.C.T. Slade, Polymer electrodes doped with heteropolymetallates and their use within solid-state supercapacitors, Synthetic Metals 139 (2003) 123.
15. L. Pan, H. Qiu, C. Dou, Y. Li, L. Pu, J. Xu, Y. Shi, Conducting polymer nanostructures: template synthesis and applications in energy storage, International Journal of Molecular Sciences 11 (2010) 2636.
16. a) Wanlu Yang, Zan Gaob, Ningning Song, Yunya Zhang, Yingchao Yang, Jun Wang, Synthesis of hollow polyaniline nano-capsules and their supercapacitor application, Journal of Power Sources 272 (2014) 915.
    b) C. Yuan, L. Zhang, L. Hou, J. Li, G. Pang, Green interfacial synthesis of two-dimensional poly(2,5-dimethoxyaniline) nanosheets as a promising electrode for high performance electrochemical capacitors, RSC Advances 4 (2014) 24773.
17. D. Li, Y. Li, Y. Feng, H. Hu, W. Feng, Hierarchical graphene oxide/polyaniline nanocomposites prepared by interfacial electrochemical polymerization for flexible solid-state supercapacitors, Journal of Materials Chemistry A 3 (2015) 2135.
18. Q.L. Hao, H.L. Wang, X.J. Yang, L.D. Lu, X. Wang, Liquid/liquid interfacial polymerization to fabricate sulfonated graphene/polyaniline nanocomposite for supercapacitors, Applied Mechanics and Materials 29–32 (2010) 1902.
19. a) D.O. Carlsson, G. Nyström, Q. Zhou, L.A. Berglund, L. Nyholm, M. Strømme, Electroactive nanofibrillated cellulose aerogel composites with tunable structural and electrochemical properties, Journal of Materials Chemistry 22 (2012) 19014.
    b) Y. Xu, Z. Sui, B. Xu, H. Duan, X. Zhang, Emulsion template synthesis of all conducting polymer aerogels with superb adsorption capacity and enhanced electrochemical capacitance, Journal of Materials Chemistry 22 (2012) 8579.
20. Y. Lu, W. He, T. Cao, H. Guo, Y. Zhang, Q. Li, Z. Shao, Y. Cui, X. Zhang, Elastic, conductive, polymeric hydrogels and sponges, Scientific Reports 4 (2014) 5792.
21. a) Y. Shi, G. Yu, Designing hierarchically nanostructured conductive polymer gels for electrochemical energy storage and conversion, Chemistry of Materials 28 (2016) 2466.
    b) M. Nazari, S. Agbolaghi, S. Abbaspoor, H. Gheybi, F. Abbasi, Arrangement of conductive rod nanobrushes via conductive–dielectric–conductive sandwiched single crystals of poly(ethylene glycol) and polyaniline block copolymers, Macromolecules 48 (2015) 8947.
22. F.G. Moore, G.L. Richmond, Integration or segregation: how do molecules behave at oil/water interfaces? Accounts of Chemical Research 41 (2008) 739.
23. Y. Xia, R. Fang, Z. Xiao, L. Ruan, R. Yan, H. Huang, C. Liang, Y. Gan, J. Zhang, X. Tao, W. Zhang, Supercritical fluid assisted biotemplating synthesis of Si–O–C microspheres from microalgae for advanced Li-ion batteries, RSC Advances 6 (2016) 69764.
24. a) M. Sarmadivaleh, A.Z. Al-Yaseri, S. Iglauer, Influence of temperature and pressure on quartz–water–CO2 contact angle and CO2–water interfacial tension, Journal of Colloid and Interface Science 441 (2015) 59.
    b) N. Shohai Kaveh, E.S.J. Rudolph, P. van Hemert, W.R. Rossen, K.H. Wolf, Wettability evaluation of a CO2/water/bentheimer sandstone system: contact angle, dissolution, and bubble size, Energy & Fuels, 28, 2014, p. 4002.

b) N.J. Hyne Petroleum: Geology, Exploration, Drilling & Production, PennWell Corporation, Tulsa, OK, USA, 2012.
c) A. Khusainova, S.M. Nielsen, H.H. Pederson, J.M. Woodley, A. Shapiro, Study of wettability of calcite surfaces using oil−brine−enzyme systems for enhanced oil recovery applications, Journal of Petroleum Science and Engineering 127 (2015) 53.
25. H. Lin, L. Feng, L. Xingang, S. Hong, X. Zhenghe, Interfacial sciences in unconventional petroleum production: from fundamentals to applications, Chemical Society Reviews 44 (2015) 5446.
26. Y. Ruiz-Morales, O.C. Mullins, Coarse-grained molecular simulations to investigate asphaltenes at the oil−water interface, Energy Fuels 29 (2015) 1597.
27. I. Benjamin, Molecular structure and dynamics at liquid-liquid interfaces, Annual Review Physical Chemistry 48 (1997) 407.
28. P. Jungwirth, B.J. Finlayson-Pitts, D.J. Tobias, Introduction: structure and chemistry at aqueous interfaces, Chemical Reviews 106 (2006) 1137.
29. J. Sjoblom, N. Aske, I.H. Auflem, O. Brandal, T.E. Havre, O. Sather, A. Westvik, E. E. Johnsen, H. Kallevik, Our current understanding of water-in-crude oil emulsions.: Recent characterization techniques and high pressure performance, Advances in Colloid and Interface Science 100 − 102 (2003) 399.
30. Y. Fan, S. Simon, S. Sjoblom, Interfacial shear rheology of asphaltenes at oil−water interface and its relation to emulsion stability: Influence of concentration, solvent aromaticity and nonionic surfactant, Colloids and Surfaces A 366 (2010) 120.
31. a) T. Hoefling, D. Stofesky, M. Reid, E. Beckman, R.J. Enick, The incorporation of a fluorinated ether functionality into a polymer or surfactant to enhance C02-solubility, Journal of Supercritical Fluids 5 (1992) 237.
b) T.A. Heofling, D.A. Newman, R.M. Enick, E. Beckman, Effect of structure on the cloud-point curves of silicone-based amphiphiles in supercritical carbon dioxide, Journal of Supercritical Fluids 6 (1993) 165.
c) T.A. Hoefling, R.R. Beitle, R.M. Enick, E.J. Beckman, Design and synthesis of highly CO2-soluble surfactants and chelating agents, Fluid Phase Equilibria 83 (1993) 203.
32. S. Zhang, Y. She, Y. Gu, Evaluation of polymers as direct thickeners for CO2 enhanced oil recovery, Journal of Chemical and Engineering Data 56 (2011) 1069.
33. J. Xu, A. Wlaschin, R.M. Enick, SPE Journal 8 (2003) 85.
34. V.H. Dalvi, V. Srinivasan, P.J. Rossky, Understanding the effectiveness of fluorocarbon ligands in dispersing nanoparticles in supercritical carbon dioxide, Journal of Physical Chemistry C 114 (2010) 15553.
35. Y. Wang, L. Hong, D. Tapriyal, I.C. Kim, I.-H. Paik, J.M. Crostwaite, A.D. Hamilton, M.C. Thies, E.J. Beckman, R.M. Enick, J.K. Johnson, Design and evaluation of nonfluorous CO2-soluble oligomers and polymers, Journal of Physical Chemistry B 113 (2009) 14971.
36. P. Raveendran, S.L. Wallen, Sugar acetates as novel, renewable CO2-philes, Journal of the American Chemical Society 124 (2002) 12590.
37. a) H. Rezaei, S. Amjad-Iranagh, H. Modarress, Self-Accumulation of uncharged polyaromatic surfactants at crude oil−water interface: a mesoscopic dpd study, Energy Fuels 30 (2016) 6626.
b) J.D. McLean, P.K. Kilpatrick, Effects of Asphaltene Aggregation in model heptane−toluene mixtures on stability of water-in-oil emulsions, Journal of Colloid and Interface Science 196 (1997) 23.
38. L.Y. Zhang, R. Lopetinsky, Z. Xu, J.H. Masliyah, Asphaltene Films at a toluene/water interface, Energy Fuels 21 (2007) 274.

39. M. Li, M. Xu, Y. Ma, Z. Wu, A.A. Christy, Interfacial film properties of asphaltenes and resins, Fuel 81 (2002) 1847.
40. Y. Mikami, Y. Liang, T. Matsuoka, E.S. Boek, Molecular dynamics simulations of asphaltenes at the oil–water interface: from nanoaggregation to thin-film formation, Energy Fuels 27 (2013) 1838.
41. C. Jian, M.R. Poopari, Q. Liu, N. Zerpa, H. Zeng, T. Tang, Reduction of water/oil interfacial tension by model asphaltenes: the governing role of surface concentration, Journal of Physical Chemistry B 120 (2016) 5646.
42. K.S. Petersen, P.L. Christensen, Phase Behavior of Petroleum Reservoir Fluids, CRC Press (Taylor and Francis Group), Boca Raton, FL, USA, 2007.
43. a) S. Majumdar, U. Saikia, D. Mahanta, Polyaniline-coated filter papers: cost effective hybrid materials for adsorption of dyes, Journal of Chemical & Engineering Data 60 (2015) 3382.
    b) V. Fasano, M. Moffa, A. Camposeo, L. Persano, D. Pisignano, Controlled atmosphere electrospinning of organic nanofibers with improved light emission and waveguiding properties, Macromolecules 48 (2015) 7803.
    c) X. Gan, H. Zhao, S. Chen, H. Yu, X. Quan, Three-dimensional porous HxTiS2 nanosheet–polyaniline nanocomposite electrodes for directly detecting trace Cu (II) ions, Analytical Chemistry 87 (2015) 5605.
44. J.P. Dickie, T.F. Yen, Macrostructures of the asphaltic fractions by various instrumental methods, Analytical Chemistry 39 (1967) 1847.
45. O.C. Mullins, The Modified Yen Model, Energy Fuels 24 (2010) 2179.
46. T.F. Headen, E.S. Boek, J. Stellbrink, U.M. Scheven, Small angle neutron scattering (sans and v-sans) study of asphaltene aggregates in crude oil, Langmuir 25 (2009) 422.
47. L. Barre, S. Simon, and T. Palermo, Solution properties of asphaltenes, Langmuir 24 (2008) 3709.
48. F. Mostowfi, K. Indo, O.C. Mullins, R. McFarlane, Asphaltene nanoaggregates studied by centrifugation, Energy & Fuels 23 (2009) 1194.
49. E.S. Boek, H.K. Ladva, J.P. Crawshaw, J.T. Padding, Deposition of colloidal asphaltene in capillary flow: experiments and mesoscopic simulation, Energy & Fuels 22 (2008) 805.
50. T.F. Headen, E.S. Boek, N.T. Skipper, Evidence for asphaltene nanoaggregation in toluene and heptane from molecular dynamics simulations, Energy & Fuels 23 (2009) 1220.
51. M. Hmoudah, N.N. Nassar, G. Vitale, A. El-Qannia, Effect of nanosized and surface-structuralmodified nano-pyroxene on adsorption of violanthrone-79, RSC Adv. 6 (2016) 64482.
52. a) L. He, F. Lin, X. Li, H. Sui, Z. Xu, Interfacial sciences in unconventional petroleum production: from fundamentals to applications, Chemical Society Reviews 44 (2015) 5446.
    b) B. Mostofian, C.M. Cai, M.D. Smith, L. Petridis, X. Cheng, C.E. Wyman, J.C. Smith, Local phase separation of co-solvents enhances pretreatment of biomass for bioenergy applications, Journal of the American Chemical Society 138 (2016) 10869.
53. J.K. Hasty, S. Ponnurangam, S. Turn, P. Somasundaran, T. Kim, D. Mahajan, Catalytic synthesis of mixed alcohols mediated with nano-MoS2 microemulsions, Fuel 164 (2016) 339.
54. a) C.M. Starks, Phase-transfer catalysis. I. Heterogeneous reactions involving anion transfer by quaternary ammonium and phosphonium salts, Journal of the American Chemical Society 93 (1971) 195.
    b) M. Makosza, Phase-transfer catalysis. A general green methodology in organic synthesis, Pure and Applied Chemistry 72 (2000) 1399.

c) B.H. Lipshutz, N.A. Isley, R. Moser, S. Ghorai, H. Leuser, B.R. Taft, Rhodium-catalyzed asymmetric 1,4-additions, in water at room temperature, with in-flask catalyst recycling, Advanced Synthesis and Catalysis 354 (2012) 3175.

55. S. Crossley, J. Faria, M. Shen, D.E. Resasco, Solid nanoparticles that catalyze biofuel upgrade reactions at the water/oil interface, Science 327 (2010) 68.
56. Y.Z. Chen, G. Cai, Y. Wang, Q. Xu, S.-H. Yu, H.-L. Jiang, Palladium nanoparticles stabilized with N-doped porous carbons derived from metal−organic frameworks for selective catalysis in biofuel upgrade: the role of catalyst wettability, Green Chemistry 18 (2016) 1212.
57. G. Huang, Q. Yang, Q. Xu, S.-H. Yu, H.-L. Jiang, Polydimethylsiloxane coating for a palladium/MOF composite: Highly improved catalytic performance by surface hydrophobization, Angewandte Chemie International Edition 55 (2016) 7379.
58. a) X. Xu, Y. Li, Y. Gong, P. Zhang, H. Li, Y. Wang, Synthesis of palladium nanoparticles supported on mesoporous N-doped carbon and their catalytic ability for biofuel upgrade, Journal of the American Chemical Society 134 (2012) 16987.
    b) Z. Zhu, H. Tan, J. Wang, S. Yu, K. Zhou, Hydrodeoxygenation of vanillin as a bio-oil model over carbonaceous microspheres-supported Pd catalysts in the aqueous phase and Pickering emulsions, Green Chemistry 16 (2014) 2636.
    c) X. Yang, Y. Liang, Y. Cheng, W. Song, X. Wang, Z. Wang, J. Qiu, Hydrodeoxygenation of vanillin over carbon nanotube-supported Ru catalysts assembled at the interfaces of emulsion droplets, Catalysis Communications 47 (2014) 28.
59. L. Liu, X. Pu, Y. Zhou, X. Wu, D. Luo, Z. Ren, Phase inversion of pickering emulsions by electrolyte for potential reversible water-in-oil drilling fluids, Energy Fuels 34 (2020) 1317.
60. a) J. Faria, M.P. Ruiz, D.E. Resasco, Phase-selective catalysis in emulsions stabilized by janus silica-nanoparticles, Advanced Synthesis & Catalysis 352 (2010) 2359.
    b) M.P. Ruiz, J. Faria, M. Shen, S. Drexler, T. Prasomsri, D.E. Resasco, Nanostructured carbon−metal oxide hybrids as amphiphilic emulsion catalysts, ChemSusChem 4 (2011) 964.
    c) S. Drexler, J. Faria, M.P. Ruiz, J.H. Harwell, D.E. Resasco, Amphiphilic nanohybrid catalysts for reactions at the water/oil interface in subsurface reservoirs, Energy & Fuels 26 (2012) 2231.
    d) P.A. Zapata, J. Faria, M.P. Ruiz, D.E. Resasco, Condensation/hydrogenation of biomass-derived oxygenates in water/oil emulsions stabilized by nanohybrid catalysts, Topics in Catalysis 55 (2012) 38.
    e) D. Shi, J.A. Faria, A.A. Rownaghi, R.L. Huhnke, D.E. Resasco, Fischer−Tropsch synthesis catalyzed by solid nanoparticles at the water/oil interface in an emulsion system, Energy & Fuels 27 (2013) 6118.
61. a) X. Xu, Y. Gong, P. Zhang, H. Li, Y. Wang, Synthesis of palladium nanoparticles supported on mesoporous N-doped carbon and their catalytic ability for biofuel upgrade, Journal of the American Chemical Society 134 (2012) 16987.
    b) C. Yu, L. Fan, J. Yang, Y. Shan, J. Qiu, Phase-reversal emulsion catalysis with CNT−TiO2 nanohybrids for the selective oxidation of benzyl alcohol, Chemistry: A European Journal 19 (2013) 16192.
    c) A.P.C. Teixeira, A.D. Purceno, A.S. Barros, B.R.S. Lemos, J.D. Ardisson, W.A.A. Macedo, E.C.O. Nassor, C.C. Amorim, F.C.C. Moura, M.G. Hernandez-Terrones, F.S.M. Portela, R.M. Lago, Amphiphilic magnetic composites based on layered vermiculite and fibrous chrysotile with carbon nanostructures: Application in catalysis, Catalysis Today 190 (2012) 133.

d) Y. He, F. Wu, X. Sun, R. Li, Y. Guo, C. Li, L. Zhang, F. Xing, W. Wang, J. Gao, ACS Appl. Mater. Interfaces 5 (11) (2013) 4843–4855. *ACS Applied Materials and Interfaces*, 2013, **5**, 4843.
62. W. Fang, Z. Fan, H. Shi, S. Wang, W. Shen, H. Xu, J.-M. Clacens, F.D.E. Campo, A. Liebensa, M. Pera-Titus, Aquivion®—carbon composites via hydrothermal carbonization: amphiphilic catalysts for solvent-free biphasic acetalization, Journal of Materials Chemistry A 4 (2016) 4380.
63. a) M.-M. Titirici, M. Antonietti, Chemistry and materials options of sustainable carbon materials made by hydrothermal carbonization, Chemical Society Reviews 39 (2010) 103.
b) C. Falco, N. Baccile, M.-M. Titirici, Morphological and structural differences between glucose, cellulose and lignocellulosic biomass derived hydrothermal carbons, Green Chemistry 13 (2011) 3273.
64. M. Pera-Titus, L. Leclercq, J.-M. Clacens, F. De Campo, V. Nardello-Rataj, Pickering Interfacial catalysis for biphasic systems: from emulsion design to green reactions, Angewandte Chemie International Edition 54 (2015) 2006.
65. a) G.F. Ghesti, J.L. de Macedo, V.C.I. Parente, J.A. Dias, S.C.L. Dias, Synthesis, characterization and reactivity of Lewis acid/surfactant cerium trisdodecylsulfate catalyst for transesterification and esterification reactions, Applied Catalysis A 355 (2009) 139.
b) F.C.G. de Mattos, J.A.D. de Souza, A.B.D. Cotrim, J.L. de Macedo, J.A. Dias, S.C. L. Dias, G.F. Ghesti, Lewis acid/surfactant rare earth trisdodecylsulfate catalysts for biodiesel production from waste cooking oil, Applied Catalysis, A 423 (2012) 1.
c) P. Gaudin, R. Jacquot, P. Marion, Y. Pouilloux, F. Jerome, Acid-catalyzed. etherication of glycerol with long-alkyl-chain alcohols, ChemSusChem 4 (2011) 719.
66. Y. Zhou, C. Jia, R. Li, X. Zhang, W. He, J. Li, B. Feng, Q. Xia, An efficient and expeditious synthesis of phytostanyl esters in a solvent-free system, European Journal of Lipid Science and Technology 114 (2012) 896.
67. J. Potier, S. Menuel, M.-H. Chambrier, L. Burylo, J.-F. Blach, P. Woisel, E. Monflier, F. Hapiot, Pickering emulsions based on supramolecular hydrogels: application to higher olefins' hydroformylation, ACS Catalysis 3 (2013) 1618.
68. a) H. Li, J. Chen, L. Hua, Y. Qiao, Y. Yu, Z. Pan, H. Yang, Z. Hou, Polyoxometalate and copolymer-functionalized ionic liquid catalyst for esterification, Pure Applied Chemistry 84 (2012) 541.
b) Y. Ding, W. Zhao, H. Hua, B. Ma, [π-C5H5N(CH2)15CH3]3[PW4O32]/H2O2/ethyl acetate/alkenes: a recyclable and environmentally benign alkenes epoxidationcatalytic system, Green Chemistry 10 (2008) 910.
69. Q. Wu, H. Wan, H. Li, H. Song, T. Chu, Bifunctional temperature-sensitive amphiphilic acidic ionic liquids for preparation of biodiesel, Catalysis Today 200 (2013) 74.
70. a) J. Li, A. Takeuchi, M. Ozawa, X. Li, K. Saigo, K. Kitazawa, C60 fullerol formation catalysed by quaternary ammonium hydroxides, Journal of Chemical Society Chemical Communications 1993 (1784) 23.
b) M. Mikawa, H. Kato, M. Okumura, M. Narazaki, Y. Kanazawa, N. Miwa, H. Shinohara, Paramagnetic water-soluble metallofullerenes having the highest relaxivity for MRI contrast agents, Bioconjugate Chemistry 12 (2001) 510.
71. K. Shimada, Y. Ohe, T. Ohguni, K. Kawano, J. Ishii, T. Nakamure, N. Shokuhin, K. Gakkaishi, Journal of the Japanese Society for Food Science and Technology 38 (1991) 16.
72. N. Joseph, J. Varghese, M.T. Sebastian, Self-assembled polyaniline nanofibers with enhanced electromagnetic shielding properties, RSC Advances 5 (2015) 20459.
73. L. Leclercq, R. Company, A. Mhlbauer, A. Mouret, J.-M. Aubry, V. Nardello-Rataj, Versatile eco-friendly pickering emulsions based on substrate/native cyclodextrin

complexes: a winning approach for solvent-free oxidations, ChemSusChem 6 (2013) 1533.

74. Juan M. Rodnguez Patino, M. Rosario Rodriguez Nino, Cecilio Carrera Sanchez, Dynamic interfacial rheology as a tool for the characterization of whey protein isolates gelation at the oil-water interface, J. Agric. Food Chem. 47 (1999) 3640–3648.
75. a) A. Sherburn, M. Platt, D.W.M. Arrigan, N.M. Boag, R.A.W. Dryfe, Selective silver ion transfer voltammetry at the polarised liquid|liquid interface, Analyst 128 (2003) 1187–1192.
    b) D.W.M. Arrigan, M. Ghita, V. Beni, Chem. Selective voltammetric detection of dopamine in the presence of ascorbate, Commun (2004) 732–733.
    c) A.M. O'Mahony, M.D. Scanlon, A. Berduque, V. Beni, D.W.M. Arrigan, E. Faggi, A. Bencini, Voltammetry of chromium(VI) at the liquid|liquid interface, Electrochem. Commun 7 (2005) 976–982.
    d) Y. Chen, Y. Yuan, M. Zhang, F. Li, P. Sun, Z. Gao, Y. Shao, Sci. China, Ser. B 47 (2004) 24–33.
    e) T. Osakai, T. Hirai, T. Wakamiya, S. Sawada, Phys. Quantitative analysis of the structure–hydrophobicity relationship for di- and tripeptides based on voltammetric measurements with an oil/water interface, Chem.Chem. Phys 8 (2006) 985–993.
    f) M. Shinshi, T. Sugihara, T. Osakai, M. Goto, Electrochemical extraction of proteins by reverse micelle formation, Langmuir 22 (2006) 5937–5944.
    g) J. Guo, S. Amemiya, Voltammetric Heparin-selective electrode based on thin liquid membrane with conducting polymer-modified solid support, Anal. Chem 78 (2006) 6893–6902.
    i) A. Berduque, M.D. Scanlon, C.J. Collins, D.W. Arrigan, M. Electrochemistry of non-redox-active poly(propylenimine) and poly(amidoamine) dendrimers at liquid – liquid interfaces, Langmuir 23 (2007) 7356–7364.
    j) G. Herzog, V. Kam, A. Berduque, D.W.M. Arrigan, Detection of food additives by voltammetry at the liquid-liquid interface, J. Agric. Food Chem 56 (2008) 4304–4310.
    k) A. Berduque, A. Sherburn, M. Ghita, R.A.W. Dryfe, D.W.M. Arrigan, Electrochemically modulated liquid – liquid extraction of ions, Anal. Chem 77 (2005) 7310–7318.
    l) A. Berduque, D.W.M. Arrigan, Selectivity in the coextraction of cation and anion by electrochemically modulated liquid – liquid extraction, Anal. Chem 78 (2006) 2717–2725.
    m) J. Courtney, J. Collins, D.W.M. Arrigan, A review of recent advances in electrochemically modulated extraction methods, Analytical and Bioanalytical Chemistry 393 (3) (2009) 835–845.
    o) J. Strutwolf, M.D. Scanlon, D.W.M. Arrigan, Electrochemical ion transfer across liquid/liquid interfaces confined within solid-state micropore arrays – simulations and experiments, The Analyst 134 (1) (2009) 148–158.
    p) D.W.M. Arrigan, Bioanalytical detection based on electrochemistry at interfaces between immiscible liquids, Analytical Letters 41 (18) (2008) 3233–3252.
76. B.M. Sischer, Y.-Y. Won, D.S. Ege, J.C.-M. Lee, F.S. Bates, D.E. Discher, D.A. Hammer, Polymersomes: tough vesicles made from Diblock copolymers, Science 284 (1999) 1143.
77. Z. Wang, M.C.M. van Oers, F.P.J.T. Rutjes, J.C.M. van Hest, Polymersome colloidosomes for enzyme catalysis in a biphasic system, Angewandte Chemie International Edition 51 (2012) 10746.
78. J. Liu, G. Lan, Y. Peng, Y. Li, C. Li, Q. Yang, Enzyme confined in silica-based nanocages for biocatalysis in a Pickering emulsion, Chemical Communications 49 (2013) 9558.

79. Y. Jiang, X. Liu, Y. Chen, L. Zhou, Y. He, L. Ma, J. Gao, Pickering emulsion stabilized by lipase-containing periodic mesoporous organosilica particles: A robust biocatalyst system for biodiesel production, Bioresources Technology 153 (2014) 278.
80. C. Hou, H. Zhu, D. Wu, Y. Li, K. Hou, Y. Jiang, Y. Li, Immobilized lipase on macroporous polystyrene modified by PAMAM-dendrimer and their enzymatic hydrolysis, Process Biochemistry 49 (2014) 244.
81. D. Kim, S. Cheong, Y. Gyong, A. Sook, W. Ryu, J.-K. Kim, J. Cho, Multicatalytic colloids with highly scalable, adjustable, and stable functionalities in organic and aqueous media, Nanoscale 8 (2016) 7000.
82. R. Chinchilla, C. Nájera, The Sonogashira Reaction: A booming methodology in synthetic organic chemistry, Chemical Reviews 107 (2007) 874.
83. J.M. Pedersen, W.R. Bowman, M.R.J. Elsegood, A.J. Fletcher, P.J. Lovell, Synthesis of ellipticine: A radical cascade protocol to aryl- and heteroaryl-annulated[b]carbazoles, Journal of Organic Chemistry 70 (2005) 10615.
84. J. Kim, J.E. Lee, J. Lee, Y. Jang, S.-W. Kim, K. An, J.H. Yu, T. Hyeon, Generalized fabrication of multifunctional nanoparticle assemblies on silica spheres, Angewandte Chemie International Edition 45 (2006) 4789.
85. W. Zhu, H. Li, X. He, Q. Zhang, H. Shu, Y. Yan, Synthesis of adipic acid catalyzed by surfactant-type peroxotungstates and peroxomolybdates, Catalysis Communications 9 (2008) 551.
86. M. Zhang, L. Wei, H. Chen, Z. Du, B.P. Binks, H. Yang, Compartmentalized droplets for continuous flow liquid–liquid interface catalysis, Journal of the American Chemical Society 138 (2016) 10173.
87. a) E. Bakker, Electroanalysis with membrane electrodes and liquid–liquid interfaces, Analytical Chemistry 88 (1) (2016) 395–413.
    b) Y. Liu, J. Strutwolf, D.W.M. Arrigan, Ion-transfer voltammetric behavior of propranolol at nanoscale liquid–liquid interface arrays, Anal. Chem. 87 (2015) 4487–4494.
    c) N. Kikkawa, L. Wang, A. Morita, Microscopic barrier mechanism of ion transport through liquid–liquid interface, J. Am. Chem. Soc. 137 (2015) 8022–8025.
    d) S. O'Sullivan, E. Alvarez de Eulate, Y.H. Yuen, E. Helmerhorst, D.W.M. Arrigan, Stripping voltammetric detection of insulin at liquid–liquid microinterfaces in the presence of bovine albumin, Analyst 138 (2013) 6192–6196.
    e) N.K. Joon, Ning He, T. Ruzgas, J. Bobacka, G. Lisak, PVC-based ion-selective electrodes with a silicone rubber outer coating with improved analytical performance, Analytical Chemistry 91 (16) (2019) 10524–10531.
    f) E. Lindner, M. Guzinski, T.A. Khan, B.D. Pendley, Reference electrodes with ionic liquid salt bridge: When will these innovative novel reference electrodes gain broad acceptance? ACS Sensors 4 (3) (2019) 549–561.
88. a) M. Inoue, K. Hashizaki, H. Taguchi, Y. Saito, Preparation and characterization of n-alkane/water emulsion stabilized by cyclodextrin, Journal of Dispersion Science & Technology 2010 (1648) 31.
    b) L. Davarpanah, F. Vahabzadeh, Formation of oil-in-water (O/W) pickering emulsions via complexation between β-cyclodextrin and selected organic solvents, Starch 64 (2012) 898.
89. J.R. Fried, Polymer Science and Technology, Prentice Hall, Upper Saddle River, NJ, USA, 2003. Chapter 10.
90. L. Feng, C. Zhu, H. Yuan, L. Liu, F. Lv, S. Wang, Conjugated polymernanoparticles: preparation, properties, functionalization and biological applications, Chemical Society Reviews 42 (2013) 6620.
91. P. Dallas, R. Zboril, A.B. Bourlinos, D. Jancik, D. Niarchos, A. Panacek, D. Petridis, Cornet-Like Phosphotriazine/diamine polymers as reductant and matrix for the

synthesis of silver nanocomposites with antimicrobial activity, Macromolecular Materials Engineering 295 (2010) 108.
92. a) S. Jeshycka, H.Y. Han, H.J. Lee, Voltammetric understanding of ionizable doxorubicin transfer reactions across liquid/liquid interfaces and sensor development, Electrochimica Acta 245 (2017) 211.
b) L. Poltorak, K. Morakchi, G. Herzog, A. Walcarius, Electrochemical characterization of liquid-liquid micro-interfaces modified with mesoporous silica, Electrochimica Acta 179 (2015) 9–215.
93. J. Clayden, S.J.M. Rowbottom, M.G. Hutchings, W.J. Ebenezer, Formation of water-soluble sulfonated azacalix[4]arenes from cyanuric chloride, Tetrahedron Letters 50 (2009) 3923.
94. T.H. Qazi, R. Rai, A.R. Boccaccini, Tissue engineering of electrically responsive tissues using polyaniline based polymers: a review, Biomaterials 35 (2014) 9068.
95. J. Chen, F. Wang, Q. Liu, J. Du, Antibacterial polymeric nanostructures for biomedical applications, Chemical Communications 50 (2014) 14482.
96. H.F. Chen, S.J. Yang, Z.H. Tsai, W.Y. Hung, T.C. Wang, K.T. Wong, 1,3,5-Triazine derivatives as new electron transport–type host materials for highly efficient green phosphorescent OLEDs, Journal of Materials Chemistry 19 (2009) 8112.

Further reading

S. Wiese, A.C. Spiess, W. Richtering, Microgel-stabilized smart emulsions for biocatalysis, Angewandte Chemie International Edition 52 (2013) 576.

W.-J. Zhou, L. Fang, Z. Fan, B. Albela, L. Bonneviot, F. De Campo, M.P. Era-Titus, J.-M. Clacens, et al., Tunable catalysts for solvent-free biphasic systems: pickering interfacial catalysts over amphiphilic silica nanoparticles, Journal of the American Chemical Society 136 (2014) 4869.

CHAPTER 7

Outlook and perspectives

Reviews and books on the synthesis and self-assembly of materials from interfaces are already available. The purpose of this book is to give a holistic presentation of the active environments and materials arising from these dual systems comprised of a polar and nonpolar interface, two phases that are practically separate entities. I hope it will inspire scientists and engineers for the implementation of new methodologies that will lead to the development of functional films, composites, and the application of emulsions and two-phase systems.

A historical overview is presented along with a section on the evolution of life and the self-assembly of hydrocarbons in aqueous-based, natural, or biological environments. Furthermore I discuss the recent speculations on alternative, nonaqueous, environments that could enable liposome assembly at cryogenic temperatures.

With respect to the above, I present the appearance and importance of the interface in living organisms, natural systems, and materials science. Various theories and speculations regarding the role of surfactants on the origin of life have arisen. However, even the simplest unicellular living organisms are too complex and their formation is beyond our understanding. Hence it is practically impossible to give a comprehensive approach on how living organisms have developed. In any case these theories present interesting insights on the self-organization of matter.

February 2020, Athens, Greece.

Generation of Polymers and Nanomaterials at Liquid-Liquid Interfaces
DOI: https://doi.org/10.1016/B978-0-12-819491-1.00011-3

Abbreviations

1D	one dimensional
2D	two dimensional
3D	three dimensional
AFM	atomic force microscopy
AMDEC	ammonium decanoate
AMPSA	2-acryloamino-2-methyl-1-propanesulfonic acid
APFO	ammonium perfluoroctanoate
BN	boron nitride
CD	cyclodextrin
CLSM	confocal laser scanning microscopy
CMC	critical micelle concentration
CNT	carbon nanotubes
Co_2(DPX)	biscobaltporphyrin, 4,5-bis[5-(2,8,13,17-tetraethyl-3,7,12,18-tetramethyl-porphyrynil)]-9,9-dimethylxanthene
CTAB	cetyltrimethylammonium bromide
CVD	chemical vapor deposition
DA	Diels Alder
DBS	dodecylbenzenesulfonate
DecMFc	decamethylferrocene
DLVO	Derjaguin–Landau–Verwey–Overbeek theory
DMFc	1,10-dimethylferrocene
DNA	deoxyribonucleic acid
DPD	dissipative particle dynamics
EDOT	3,4-ethylenedioxythiophene
ISC	intersystem crossing
LDH	layered double hydroxide
MD	molecular dynamics
MRI	magnetic resonance imaging
MW	molecular weight
NADP	nicotinamide adenine dinecluotide phosphate
O/W	oil in water
ORR	oxygen reduction reactions
PA	polyamide
PANI	polyaniline
PBD	polybutadiene
PC	polycarbonate
PE	polyethylene
PEG	polyethylene glycol
PMMA	polymethyl methacrylate
POM	polyoxometalates
PP	polypropylene
PPy	polypyrrole
PQS	polyethylene glycol ubiquinol sebacate

PS	polystyrene
PU	polyurethane
RNA	ribonucleic acid
SAXS	small-angle X-ray scattering
$ScCO_2$	supercritical carbon dioxide
SDS	sodium dodecyl sulfate
SEM	scanning electron microscopy
SERS	surface enhanced Raman scattering
STEM	scanning transmission electron microscopy
SWCNT	single wall carbon nanotubes
Sβ-CD	sulfonated β-cyclodextrin
Tg	glass transition temperature
UV	ultraviolet
W/O	water in oil
XRD	X ray diffraction

Index

Note: Page numbers followed by "*f*" and "*t*" refer to figures and tables, respectively.

1,2-Dichlorobenzene, 75, 92–94, 143
1,10-Dimethylferrocene (DMFc), 72, 74–75
2-Acryloamino-2-methyl-1-propanesulfonic acid (AMPSA), 163–164
3,4-Ethylenedioxythiophene (EDOT), 165, 182–183
3M Novec, 143

A

Absorb
  absorbance, 26–27, 61, 65–69, 71, 134, 148–149
  absorbed, 4, 28, 51–52, 136, 143–144
Absorption, 53–54, 59, 62–65, 134, 136, 139–141, 140*f*, 155–156, 165
  kinetics, 63
Acceptor, 7
Acetalization reaction, 176
Acetate, 10*f*, 38–39, 90*f*, 167
Acetonitrile, 14–15, 76–77
Acid, 109, 113*f*
Acidic, 102–103, 105–107, 163, 167, 175*f*
  aqueous layer, 163
Acrylonitrile, 15
Activation
  energy, 26–27, 110–111, 137–139, 171–172
Active oxygen, 181–182
Additives, 49–52, 167, 179
Adenine, 7–9, 27*f*
Adhesive, 109
Adsorbed, 59, 95, 98, 103–104, 139
  adsorption, 45, 58*f*, 102–103, 107, 161–162
Affinity, 104, 107, 139–141, 155–156, 177–178
Aggregated, 14–15, 102–103
Aggregation, 33–34, 41, 56–57, 62–65, 179
Air stability, 104
Air–aqueous interface, 98
Air–solid interface, 29–30
Air–solution interface, 53–54
Air–water, 49–50, 58*f*, 59, 65, 114–116, 133–134, 152–155, 179
Aldehyde, 119, 173–175
Aliphatic, 56*f*
  chain, 28–29, 41*f*, 99, 168, 170–171, 181–182
Alkali, 25–26, 133
  metal ion, 77, 110
Alkyl halide, 7–9, 9*f*
Alkyl, 55–56, 92–94
Allotrope, 13, 116–117
α-position, 100–101
α-tocopherol, 7–9
Alumina, 123, 176
Amine, 103–104, 117, 119, 122–123, 131–132, 149, 181–182
  group, 96–97, 146, 147*f*, 163
Ammonia, 2–6, 106–108
  gas, 105–106
Ammonium chloride, 107–108, 179
Ammonium decanoate (AMDEC), 53–54
Ammonium perchlorate, 105–106
Ammonium perfluorooctanoate (APFO), 53–54
Ammonium peroxydisulfate, 98
Ammonium persulfate, 101–102
Amorphous, 108–110, 121
Amphiphilic
  behaviour, 76
  magnetic nanoparticles, 75–76
  molecule, 1, 6–15, 33–34, 50–52, 55–56, 61, 75–77, 98–99, 114–116, 121
Angle-dependent extinction spectra, 65–67
Anhydrous, 16, 17*f*
Aniline, 96–101, 151, 163–164, 182–183

Anilinium cation, 98, 163
Anion
anionic, 53–54, 177–178
surfactants, 12
Anisotropy, 64
anisotropically shaped, 31*f*, 98
Anthracene, 114–116, 131–132, 152–153, 155
Antibacterial, 56–57, 102–103, 113–114, 122–123
Anticancer drug, 181
Aquatic environment, 6, 13–14
Aqueous
media/medium, 72, 91–92, 141–142
–oil interface, 25–26
phase, 7, 28–29, 31–32, 39, 41, 44, 51–52, 51*f*, 52*f*, 59, 61–63, 69–73, 70*f*, 78*f*, 92–94, 96–99, 97*f*, 104, 118, 139–141, 146–148, 147*f*, 163–164, 173–175, 177*f*
redox couple, 73*f*
Aquivion, 176
Aromatic, 56*f*, 98, 110, 112–114, 122–123, 150*f*, 163, 167–171, 170*f*
Array, 7, 31–32, 49, 60, 79–80, 116–117, 153–154, 182–183
Artificial photosynthesis, 12–13, 13*f*
Asphaltene, 167–171
Asymptotic law, 64
Atomic force microscopy (AFM), 45, 55–56, 67, 74
Atom transfer radical polymerization, 61–62, 103–104

B

Backbone, 96, 98, 109–111, 122–123, 146, 151–152, 154–156
Bandgap, 110, 112–113
Base detection, 108
Base, 104–107, 131–132, 141, 155–156, 171
sensor, 107–108
Bead, 170
imaging, 170*f*
Behenic acid, 16–17
Bentheimer sandstone, 166
Bentonite clay, 16–17
Benzene, 91–92, 148*f*
Benzenoid ring, 170*f*
Bilayer, 6–7, 8*f*, 10–11, 14–15, 28, 41, 50–51, 51*f*, 55–56, 56*f*, 76–77
Bimetallic, 162
Binder
binding energy, 67–68
Biodegradable polymer, 64
Biodiesel, 175*f*
Bioenergetics, 28
Biofuel upgrading, 173–176
Biological, 1–2, 4, 4*f*, 6–7, 12–13, 16–18, 23–29, 49–50, 55–56, 76–79, 165–166, 179–181, 193
Biology, 10–12, 34, 59–60
Biomass, 171–175
Biomedical, 55–56, 64, 68, 101–103, 113, 132, 183
Biomimetics, 17–18, 23, 29, 49–50, 61–62, 75–77, 76*f*, 78*f*, 79–80, 117
Bio-oil, 173–175
Biphase oxidation, 181–182
Bis(diphenyl phosphoranylidene) ammonium tetrakis (4-chlorophenyl) borate, 74
Biscobaltporphyrin, 4,5-bis-[5-(2,8,13,17-tetraethyl-3,7,12,18-tetramethylporphyrynyl)]-9,9-dimethyxanthene {Co_2(PDX)}, 75
Bitumen, 16, 17*f*, 171
Block copolymer, 52, 53*f*, 54–55, 78–79, 167
Block terpolymer, 61–63
Blue morpho, 78–79
Blue transmittance, 63–64
Boiling point, 2–3, 44–45
Boltzmann constant, 24–25, 41–42
Bond, 30–31, 35, 52*f*, 103–104, 117, 119, 152–154, 170–172, 179
Boron fluoride, 38–39
Boron nitride, 91, 152–154
Boron trifluoride, 38–39
Bovine albumin, 77

Bragg reflection, 78–79
Brewster angle microscopy, 67
Brownian motion, 63
Bulk
concentration, 25, 168–169
phase, 168–171
Butane, 16–17

C

Calcium, 60, 64–65
Camphor sulfonic acid, 106–107, 110, 163–164
Candida Antarctica lipase B, 182
Capacitance, 108, 151–152, 155–156, 163–165
Capillary, 32, 37–42, 65, 166
Carbon
black, 162
composite, 162, 176
dioxide, 2, 4, 34–35, 36*f*, 41–42, 165, 167
advancing angle, 37–38
bubble, 25–26, 37–38, 37*f*
reduction, 12–13, 13*f*
sequestration, 37–38
–water, 37–38
–quartz system, 37–38
electrode, 76–77
monoxide, 38–39
nanotube, 13, 72, 73*f*, 92–94, 100*f*, 103–104, 132–133, 132*f*
Carbonic anhydrase enzyme, 11*f*
Carbonization, 155, 176
Catalysis, 7–9, 23, 116–118, 121, 172–173, 174*t*, 176
Catalyst
deactivation, 176
Catalytic
activity, 28–29, 72–73, 76–77, 161–162, 172–175, 178*f*, 181
conversion, 16–17, 173–175, 178
cracking, 16
cycle, 181–182
effect, 172, 179–181
performance, 181
reaction, 9–10, 161, 172–175, 178, 181–182
synthesis, 172
Catalytically active, 173–175
Catanionics, 12, 53–54
Cation
anion surfactant, 12
Cationic, 10–11, 16, 53–54, 177–178
Cell depolarization, 7
Cell formation, 3–4, 6, 13–15
Cell imaging, 55–56, 131–132
Cell membrane, 6–12, 14–15, 17–18, 54
Cell respiration chain, 76–77
Celluloid material, 89
Cellulose, 65, 78–79, 89, 107–108, 165, 171–172, 176
Centrifugation, 65–67, 169
Cetyltrimethylammonium bromide (CTAB), 9*f*, 12, 49–50, 67, 98–99
Chain length, 12, 103–104
Charge hopping, 111
Charge-induced repulsion, 60
Charge-separated, 7, 34
Charge separation, 7, 164
Charge-stabilised nanoparticle, 64–65
Charge transfer, 24–25, 72–73, 132–133, 164
Charge transport, 110
Chemical conversion, 171–172
Chemical energy, 12
Chemical environment, 2, 10
Chemical flooding, 166
Chemical industry, 90
Chemical polymerization, 98
Chemical reaction, 4, 7–10, 27*f*, 28, 76
Chemical sensor, 105, 182
Chemical species, 104
Chemical structure, 89, 90*f*, 91, 91*f*, 134, 135*f*, 172*f*
Chemical vapour deposition (CVD), 91, 123, 152–153
Chemiluminescence, 11–12
Chiral, 79, 101, 182
Chlorine, 38–39, 64–65, 107, 113, 115*f*
Chloroform, 42–44, 43*f*, 95, 97*f*, 98, 106–109, 113*f*, 143–144, 144*f*, 145*f*, 148, 177–181
Chlorophyll, 26–29

Clay, 16–17
Cluster, 60, 169, 177–178
Coal, 16
Coalescence, 16, 41–42
Coarse grain bead, 170
Cobalt, 11–12, 107
  phthalocyanine, 76–77
Codeposition, 162
Coefficient, 24–25, 39–40, 63, 95, 148–149
Colloid, 31*f*, 42–44, 59–60, 68, 179–181
  and interface science, 60
Colloidal
  nanoparticle, 179–181
  polymer mixture, 41–42
  suspension, 41–42
  system, 29–30, 64–65
Colorimetric sensor, 107–108
Color, 28–29, 68, 75–76, 138*f*, 149
Column chromatography, 58*f*, 148–149
Concentration, 9–10, 24–25, 24*f*, 38–39, 49–50, 53*t*, 54–55, 60–61, 64–67, 69–70, 71*f*, 78*f*, 92–95, 98–99, 101–105, 108, 121, 168–169, 171–175
Condensation, 91–95
  reaction, 89–90, 121
Conductance, 70, 109–111
Conducting electrical material, 70
Conductive, 72–73, 90–92, 96, 98, 105, 141–142, 161–162, 182–183
Conductivity, 75, 79–80, 90, 96, 104–105, 107, 110–112, 155–156, 163–165
Confocal laser scanning microscopy (CLSM), 41–42
Conformational alignment, 106–107
Conjugate
  conjugated, 28–29, 91, 91*f*, 150*f*
  conjugation, 110, 151–152
Construction, 12, 29, 31–32, 76–77, 79–80, 104
Contact angle, 25–26, 32, 37–38, 59, 63–64, 118–119, 155–156, 171, 175–176
Continuous phase, 25–26, 30
Conversion of energy, 161–162
Cooled, 31–32
Copolymer, 53*f*, 54–55, 61–62, 78
Copper, 9, 107, 110, 116–117, 165, 176, 181
Cost, 172
Coupling, 7–9, 26–27, 56–57, 154–155
  reaction, 9, 72, 116–117, 181
Cracking, 16
Critical micelle concentration (CMC), 49–50, 52, 53*f*, 54, 98–99, 146–148
Critical temperature, 34–35
Crosslinked, 148*f*
Crude oil, 37–38, 166–168, 171
Cryogenic solvent, 14
Crystal
  coalescence, 16
  growth, 16
  structure, 76
  crystalline, 50–51, 108–109, 154–155, 182–183
    nanoparticle, 109
    polymers, 109, 119
  crystallinity, 59, 90, 109–111, 121
  crystallization, 16, 54–55, 108–110
Current
  density, 4, 77, 163–166
  flow, 26, 72–73
Cyclic voltammetry, 28–29, 72, 97
Cycloaddition, 7–10, 9*f*, 152–153
Cyclodextrin (CD), 65, 101, 172–173, 174*t*
Cyclohexane, 62–63, 95
Cylindrical, 32–33, 61
Cytochrome c oxidase, 76–77

D

Deactivation, 176
Decamethylferrocene (DecMFc), 28–29, 74, 77
Decane, 31–32, 31*f*, 59
Decomposition, 17, 17*f*, 67, 105–106, 171–172
Dedoping, 106–108
Deformation, 32–33, 40, 42–44, 78, 94*f*, 120*f*
Degree(s) of polymerisation, 54–55

Delocalisation, 110, 112–113, 123, 164
Dendrimer, 72, 179–181
Dense packing, 181
Density, 4, 23, 37–38, 38*t*, 41–42, 53–54, 56–57, 99–100, 113–114, 168–169, 171
Deoxyribonucleic acid (DNA), 5–6, 5*f*, 49, 59–60, 68, 133, 154–155
Depletion layer, 51–52
Deposition, 28–29, 64, 67, 72, 74, 161–162
Deprotonating, 104
Dejaguin–Landau–Verway–Overbeek theory (DLVO), 64–65
Diamagnetic, 92–94, 112–113
Diameter, 15*f*, 55–57, 59, 61, 67–69, 71*f*, 94–95, 99, 101–102, 106–107, 117–118, 151, 163–164
Diamond, 13
Dichlorobenzene, 92–94
Dichloroethane, 74
Dichloromethane, 42–44, 97*f*, 98, 113–114, 116–117, 143, 148, 153–155, 164
Dielectric
  constant, 34
  liquid, 32–33
  spectra, 111–112
Diels–Alder (DA), 7–10
Differential scanning calorimetry, 51–52
Diffuse, 98
Diffusion
  coefficient, 24–25, 63
  layer, 25, 26*f*
Dipolar, 35, 71
Dipole
  approximation, 71
  moment, 34, 38–39
Disperse, 102–103, 165
Dissipative particle dynamics (DPD), 167–168, 170–171
Dissociation, 6, 164, 179
Dissolution, 16, 31–32, 44, 74
Distribution, 3*f*, 35, 36*f*, 40–42, 56–57, 90–92, 101–103, 106–107, 166, 169–170
Dodecylbenzenesulfonic acid (DBS), 10*f*, 11–12
Dopant, 99–100, 106–107, 163
Doping, 101, 106–107, 109–111, 113*f*, 163–164
Double-wall ring interfacial rheometer, 65
Double-tail phosphate surfactant, 41–42
Doughnut shape, 143, 144*f*
Drop kicking, 25–26
Drop method, 32, 62–63, 168–169
Drude model, 111–112
Drug, 55–56, 118, 141, 181
Drying, 34–35, 165
D-spacing, 108–109
DuPont, 89–90
Dye, 41–42, 118, 136, 139–141, 182
  absorption, 165
Dynamic, 49–50
  light scattering, 55–56, 58*f*, 118

E

Einstein equation, 24–25
Elasticity, 6, 14
Electric
  discharge, 4, 13
  field, 7, 8*f*, 25, 32–33, 38–39, 67, 105, 136
  polarization, 70–71
Electrical
  conductance, 70
  conductivity, 90, 94–95, 104–105, 151–152
  properties, 101
  resistance, 90, 104–107
Electro density, 35
Electrochemical
  applications, 72–73, 79–80
  capacitance, 165
  deposition, 74
  impedance spectroscopy, 72
  properties, 55–56, 146
Electrochemistry, 72, 77, 79–80, 179
Electrochromic, 183
Electrode, 5–6, 76–77, 90, 141–142, 161–162, 164–165, 182
Electrodynamics, 68–69

Electrolyte, 60, 65–67, 70*f*, 72–75, 73*f*, 142, 163–164, 179
Electromagnetic field, 28
Electromagnetic radiation, 4, 12
Electron
  acceptor, 110, 161–162
  charge, 2
  cloud, 68–69, 123
  conductivity, 75
  delocalization, 110, 112–113
  density, 35
  donor, 7, 74–75, 77, 110
  hole-pair, 164
  hopping, 110, 164
  lattice coupling, 110
  microscopy, 11–12, 74, 118, 171–172
  paramagnetic resonance, 105
  rich palladium catalyst, 7–9
  reduction, 75–77
  spin resonance, 11–12
  Transfer, 23, 26–29, 72–73, 164, 182–183
Electronic
  properties, 90, 96
  transport, 90, 96, 106–107, 109–113, 122–123, 141–142, 154–156, 183
Electrostatic repulsion, 64–65
Electrowetting, 32–33
Elongated, 3–4, 54, 96–98, 109
Emeraldine base, 104–106
Emeraldine salt, 98, 104–106, 109
Emission, 11–12, 68, 131–136, 144–146, 149, 154
Emulsion
  formation, 172
  stabilization, 62*f*
Energy
  applications, 78*f*, 123, 161–166
  consuming, 161–162
  conversion, 17–18, 26, 121, 161–162
  density, 108, 164–165
  harvesting, 161–162
  interface, 72
  storage, 23, 90, 116–117, 161–162, 165–166, 182–183
  transfer, 17–18, 137
  ultraviolet radiation, 4
Engineering, 17–18, 50, 96, 183
Enhanced oil recovery, 45, 90, 161, 166–167, 171
Enthalpy, 35
  enthalpic stabilization, 33–34
Entropy
  entropically, 33–34
  environment, 35
  environmentally friendly, 34
Enzyme, 10–11, 11*f*, 76–77, 92–94, 118, 172, 179–182
Epoxide, 178, 182
Equilibration, 70–71, 99–100
Equilibrium position, 67
Esterification, 173–176, 179–181
Ethanol, 25–26, 114–116, 179–181
Eucalyptus urophylla, 107–108
Eucalyptus grandis, 107–108
Evaporation, 177
Excitation maxima, 144–146
Excitation dependent, 139*f*, 143–144, 145*f*, 148–149, 148*f*, 150*f*
Exfoliation, 104, 152–155
Exposure, 104–108
External electric field, 32–33
Extraction, 7, 34, 167, 171–172, 179, 183

F

Fabrication, 29, 31*f*, 67–68, 71, 77–80, 95, 102–103, 164–165
Fenton, 11–12
Fermi level, 70–71
Ferric chloride, 28
Ferrocene, 7, 27–28, 76–77
Fibrillar, 96–97, 106–107, 163–165
Fick, 23–25
Film
  formation, 65
  morphology, 107–108
  thickness, 69, 106–107
Flame, 89–90, 114, 116*f*
Flood, 35–37
Flow, 1–2, 6–7, 26, 72–73, 77, 182
Fluid, 10–11, 34–37, 76, 165–166
Fluorescein, 10*f*, 131–132, 139–141

Fluorescence
imaging, 173–175
labeling, 173–175
quenching, 133, 136–139
spectra, 26–27
spectroscopy, 7, 11–12
Fluorescent
light, 41–42, 139–141, 144–146
properties, 41–42, 149, 155–156
Fluorinated, 44–45, 167
Fluorine, 41–42, 131–132
atom, 35–37
Fluoroacrylate, 167
Fluorocarbon
surfactant, 41–42, 53–54
water, 29–30, 34, 147*f*
Folding, 11*f*, 59–60, 117
Force, 30–32, 40, 64, 109
Formation, 3–6, 11–13, 16–17, 25–26, 44, 49–56, 95, 104–105, 110–111, 121, 137, 146, 167, 181
Fourier transform infrared spectroscopy (FTIR), 94, 151–152, 175–176
Franck–Condon, 26–27
Free energy, 26–27, 60, 63–64, 66*f*
Free-radical, 16–17
polymerization, 118, 119*f*
Fuel, 13*f*, 72–73, 108, 161–162
Fullerene, 7, 49, 55–57, 57*f*, 58*f*, 112–113, 161–162
Fused, 170, 170*f*
Fusion, 56–57

G

Gadolinium endohedral metallofullerenes, 177, 177*f*
Gas
field, 166
industry, 166–167
phase, 42–44
Gel, 31–32, 34–35, 59, 79, 165
Gellan
concentration, 64
water system, 31–32
Genus, 143, 144*f*
Gibbs, 43*f*, 60
Glass
matrix, 68
transition temperature, 54–55
Gold
alloys, 68
chloride, 101–102
concentration, 71*f*
layer, 163
nanoparticle, 56–57, 57*f*, 58*f*, 60, 64–71, 71*f*, 74, 101–102
plasmonic nanoparticles, 56–57
reflectance, 63–64
substrate, 163
Good–Girifalco, 31
Graphene, 72, 74, 79–80, 91, 104, 112–113, 152–153, 161–162, 164–165, 183
oxide, 108, 164–165
Graphite, 13, 72–73, 104, 110, 112–113, 165–166, 177
Green laser, 71*f*
Greenhouse gas, 165
Growth, 25, 39, 56–57, 109–110, 151
Gyroid, 50, 78–80

H

Hagen–Rubens approximation, 111–112
Hansen thickness, 39–40
Heat
resistant, 89
transfer, 172
Heptane, 25–26, 39, 67, 70, 71*f*, 169–170
Heterogeneous, 28–29, 72–73, 75, 90
Hexacarbonyl molybdenum, 172
Hexadecane, 118, 172
Hexagonal, 10–11, 54–55, 79, 101–102, 116–117, 170*f*
Hexamethylenediamine, 91–92
Hexane, 14–15, 68, 179–181
Homeostasis, 2
Homogenous
morphology, 110–111
reaction, 99
homogeneity, 44–45
Hop, 110
hopping, 110–111, 164
Humidity, 104–105, 141–142

Hybrid, 12–13, 13*f*, 65–67, 122–123, 152, 154–155, 173–175
Hybridized, 57*f*, 112–113, 116–117, 153–154
Hydration, 35–37, 65
Hydrazine, 105–107
Hydrocarbon
  chain, 10–11, 17, 41–42, 52
  formation, 16
Hydrochloric acid, 99, 105–109
Hydrodeoxygenation, 173–175
Hydrogen
  bonding, 7, 25–26, 30–31, 38–39, 45, 67, 117
  evolution, 72–73, 75
  peroxide, 11–12, 75–77, 177–178
  storage, 72–73
Hydrogenation, 173–176
Hydrolysis, 179–181
Hydrophilic, 1, 30*f*, 50–52, 55–56, 62*f*, 63–64, 76–77, 118, 155–156, 176, 182–183
Hydrophilicity, 104
Hydrophobic
  chain, 33, 52
  effect, 7, 54–55
  group, 179–181
  tail, 41–42, 50–52
Hydrosilylated, 76
Hydrothermal carbonization, 176
Hydroxyl groups, 38–39, 101, 177*f*
Hydroxyl radical, 11–12
Hysteresis, 112

I

Immiscible, 2, 23, 28–34, 40, 44–45, 49, 51–52, 70–73, 73*f*, 75, 113*f*, 118–119, 123, 179, 182–183
In situ, 44–45, 60, 62–63, 62*f*, 72, 94, 94*f*, 162
In vivo, 75, 177
Inert-gas, 166
Infrared
  electromagnetic radiation, 4
  spectroscopy, 35
Inhomogenous
  entity, 39–40
  film, 69
  mass distribution, 40
Initiation, 4, 18, 109, 113*f*
Initiator, 97–98, 101–102, 118, 119*f*, 163–164
Injection, 118, 143–144, 145*f* 166
Inorganic
  acids, 98–99, 110–111
  biochemical, 12–13
  electrolytes, 70*f*
  matter, 4*f*
  nanoparticle, 25
Insoluble
  film, 143
  oil, 178
Insulator, 90, 96, 111
  insulating, 70, 99, 105–107, 110
Intercalated, 11–12
Interface, 12, 24–25, 28–29, 39–42, 49, 91–92, 96–97, 139–141, 155–156, 169–170, 182–183, 193
Interfacial
  area, 33, 122–123, 172–175, 178–181
  assembly, 56–57, 71, 74*f*
  boundary, 41–42
  catalyst, 72–73, 175*f*, 176
  catalytic activity, 178*f*
  deposition, 72
  electrochemistry, 77, 79–80
  energy, 61–64, 66*f*
  film, 49, 63–64, 70–71, 104, 121
  functionalization, 177
  gold nanoparticle film, 70–71
  mirror, 70–71, 79–80
  polarity, 39
  polymerisation, 90–102, 92*f*, 104, 106–111, 113–114, 117–119, 122–123, 144*f*, 147*f*, 148–149, 162–165, 182–183
  process, 45, 75
  reaction, 98–101, 153–154, 182–183
  system, 25–26, 39, 44, 61, 72–73, 78*f*, 91–92, 96–97, 143, 155, 166–167, 171, 177–178
  tension, 7–9, 23, 29–34, 37–45, 38*t*, 51–52, 60–67, 95, 99–100, 118, 143, 144*f*, 155–156, 167–170
  thickness, 40

Intermetallic, 161–162, 173–175
Intermolecular force, 30–31
Internal reflection, 39, 52, 139–141
Interparticle distance, 64–67
Interparticle separation, 66*f*
Interstellar media, 55–56
Intracellular space, 11–12
Intersystem crossing, 136–139
Intrachain interactions, 110
Intramolecular, 40*f*, 134–136, 179
  charge transfer, 24–25
Ion
  adsorption
  exchange, 165
  flow, 1–2, 6–7, 77
  ionic
    liquid, 32–33, 44–45, 72–73, 171–172, 176
    strength, 6, 49, 53–54, 60, 66*f*
    surfactant, 53–54
    transfer, 77
  ionizing radiation, 4
Iron
  nitrate, 98, 100–101, 162
  oxide, 63–64, 72, 75–76, 176, 181
  transfer, 42–44, 72–73, 76–77
    voltammetry, 70–71
  transport, 7, 182
Irradiation, 71*f*, 137, 150*f*, 155, 164
Irreversible adsorption, 63
Isolate, 172, 180*f*
Isopropanol, 12–13, 31–32, 162, 164
Isotropic, 99

J

Janus
  nanoparticle, 61–63
  particle, 52, 61–63, 79–80, 118

K

Kapton, 89
Kinetic, 61, 63, 109–110

L

L-a-dimyristoyl-phosphatidic acid, 98
Lamellae, 1, 49–50, 78
Laser, 41–44, 139–141, 149, 150*f*
Lauryl alcohol, 176
Layer, 14–15, 49–50, 53–54, 154, 181
Layer by layer, 72, 117–118, 179–181
Layered double hydroxides (LDH), 11–12
Lecithin, 41, 51–52
  molecule, 41, 41*f*, 51–52
Length, 10–12, 17, 32–34, 41–42, 53–54, 56*f*, 65, 103–104
Leucoemeraldine, 105
Lewis acid, 34, 119, 176
Lewis base, 34, 167
Ligand, 63–64, 66*f*, 67–68, 71–72, 79–80, 113–114, 115*f*, 153–154, 167
  exchange process, 179–181
Light
  emitting, 41–42, 91–92, 99–100, 122–123, 131
  harvesting, 1–2, 153–154, 161
  microscopy, 41
  radiation, 4
  reflection, 70*f*
  scattering, 41–42
Lightning, 4–6
Lignin, 171–175
Linear, 96, 99, 104–105, 117, 146, 151–152
Lipid
  bilayer, 10–11, 14, 51*f*
  membrane, 1–2, 28, 98
Lipophilic, 9, 181–182
  reducing agent, 72–73
Liquid
  carbon dioxide, 35–39
  crystal templating, 79
  crystalline, 50–51
  disordered, 6
  exfoliation, 154
  gas boundary, 35
  gas interface, 41–42
  gas phase interface, 42–44
  liquid boundary, 39–44, 75, 96–98
  liquid interface, 18, 23, 65, 70, 74, 76–77, 91, 98, 123, 131, 163, 167, 173–175
  liquid interfacial precipitation, 56–57
  methane, 13–15

Liquid (*Continued*)
phase, 30, 32, 72–73, 118
surface, 57, 59, 114–116
Lithium batteries, 164–166
ion batteries, 165–166
Living cell, 4*f*, 6
Living matter, 17
Living organisms, 1, 5–15, 18, 179–181, 193
Loss modulus, 60
Low concentration, 24–25, 168–169
Low conductivity, 90, 96

M

Macromolecular
chain, 108–110, 151–152
network, 110
Macroscopic, 74, 114–116
thermodynamics, 40
Magnesium chloride, 60
Magnesium ion, 28–29
Magnetic
behaviour, 112
field, 76
hysteresis, 112
properties, 110–113, 141–142
resonance imaging, 177
magnetisation, 112
Marcus, Rudolph, 26–27, 27*f*
theory, 26
Mass
diffusion, 24–25, 40, 98
transfer, 16, 23–29, 179
Mechanical properties, 3–4, 6, 14–15, 67–68, 92–95, 113–114, 122–123, 155–156
Mechanical stability, 165
Melt
melting, 35
point, 2–3
Membrane potential, 7
Membrane thickness, 54–55
Mesoporous, 10–11, 76, 79, 179–181
Metal
cation, 25–26, 74, 100–102, 107, 132–133
deposition, 74
to insulator, 111
insulator transition, 111–112, 113*f*
ion, 11–12, 77, 101–103, 110, 133
ligand, 153–154
nanoparticle, 101–102
organic, 183
oxide, 177, 179–181
precursor, 74
metallic
behaviour, 111–112
film, 71
polymer, 111–112
Methane, 2–6, 12–15, 44
Methanol
oxidation, 161–162
solution, 28–29
Methanosarcina barkeri, 12–13
Micellar
assemblies, 18, 53–54, 79–80
catalysis, 7–9, 174*t*
polymerization, 99
rate constant, 9–10
reaction, 9–10
Micelle
formation, 50–54, 147*f*
structures, 50–51
Microemulsion, 23, 34, 102–103, 174*t*
Microscopy, 11–12, 41–42, 67, 74, 116*f*, 118, 141, 171–172
Mineral, 2, 16–17, 45, 67
Mirror
formation, 69
surface, 69
Miscibility, 35–37, 103–104, 171
Mixed, 44–45, 57, 65, 68, 97
Mixture, 3–4, 53–54, 68, 92–95, 162, 167–168, 171–176
Molecular dynamics, 14–15, 35–38, 44, 67–68, 117, 167–168, 182
Molecular machine, 6
Molecular mechanics, 167–168
Molecular orbital, 110, 138*f*, 170
Molecular oxygen, 76–77
Molecular recognition, 7–9, 25–26, 76–77
Molecular species, 179

Molecular structure, 41*f*, 42–44, 53–54, 111, 137–139, 170, 170*f*
Molecular weight, 7–9, 91–94, 98, 104, 139–141, 176, 179
Molten, 11*f*
Monolayer, 51–52, 67, 70–71, 70*f*, 74, 98, 121, 161–166
Monomer, 25, 35, 59, 89–90, 96–97, 100–101, 114–116, 119, 121–122, 143, 179, 181–183
Monte Carlo, 51–52, 167–168
Montmorillonite, 16
Morphology, 64, 95–97, 99, 106–111, 113–114, 117, 139, 151–152
Motion, 25–26, 41–42, 63
Mott relationship, 111
Mott temperature, 111
Multilayered graphite, 110
Multi-walled carbon nanotubes (MWCNT), 164, 176

N

Nanocomposite, 56–57, 92–95, 98, 102–103, 143–144, 164–165, 183
Nanocrystalline, 28, 121–122
Nanoelectronics, 90, 182–183
Nanofibrillar, 96–97
Nanomachine, 6, 49, 76–77
Nanoparticle, 1–2, 26*f*, 28–29, 49, 70*f*, 131–132, 161–162, 166–172
Nanoscale, 59–61, 75–77, 79, 178
Nanosheet, 116–117, 122–123, 133, 151–155
Nanostructure, 6, 29, 76, 78, 98–100, 141–142, 182–183
Narrow size distribution, 101–102, 106–107
Natural gas, 1–2
Natural oil, 16–17
Natural rubber, 89–90
Needle shaped, 98–99
Neutron
  reflectivity, 49–50, 53–54
  scattering, 51–54
n-Hexane, 9–10, 16
Nicotinamide adenine dinucleotide phosphate (NADP), 27*f*
Nitric acid, 89, 162
Nitrogen, 1–2, 4, 6, 14–15, 38–39, 101–102, 116–119, 120*f*, 168
  oxide, 108
Nitromethane–hexane system, 163–164
Noble metals, 68–69, 74, 79–80, 100–103, 122–123, 161–163, 173–175
Nomex, 89–90
Nonaqueous, 1–2, 13–15, 193
Nonflammable, 34–37, 167
Nonionic surfactant, 172
Nonpolar, 9–10, 41*f*, 55–56, 96–97, 170, 193
  solvent, 15*f*, 34, 100–101
Nontoxic, 34, 167
Nuclear magnetic resonance spectroscopy, 51–52, 143–144
Nuclear waste treatment, 75
Nucleation, 16, 97–98
Nylon, 89–95
  matrix, 92–94

O

Octane, 35–37, 182
Oil
  aqueous interface, 170–171
  extraction, 167
  formation, 16–17
  -in-water, 30, 30*f*, 61–62, 118, 175–176
  industries, 166–167, 183
  phase, 64, 104, 118, 163–164, 168–170, 182
  pyrolysis, 173–175
  recovery, 35–37, 45, 90, 141, 161, 166–167, 171
  resources, 161, 171
  tracing, 131–132
  viscosity, 166
  water, 29–30
    interface, 180*f*
      tension, 167
Olefin, 9, 167–168, 177–178
Oleic acid, 52*f*, 72
Oligomeric, 143–144, 145*f*, 151–152
Olive oil, 179–181
One-dimensional, 76, 96

One-dimensional (*Continued*)
conductor, 111
One-phase polymerization, 97, 106–107, 183
One-phase solution, 53–54
One-phase synthesis, 97
One-phase system, 97, 153–154, 167–168
Optical
microscopy, 114–116
properties, 64, 68–69, 71, 79–80, 96, 104–105, 132, 143–144, 146–148, 151, 154–155
reflectivity spectrum, 76
spectra, 143–144
spectroscopy, 26–27, 45, 151
trapping, 42–44
Order–disorder transition, 112–113
Organic
acid, 16–17, 76–77, 98, 155–156
aqueous interface, 74*f*, 97
environments, 2–4
hydrocarbon, 7
inorganic, 49, 54–59, 122–123, 152
layer, 163
media, 72, 76–77, 181–182
medium, 23, 39, 99–100
phase, 28–29, 31–32, 39, 41–44, 43*f*, 52, 61–63, 74, 95–98, 97*f*, 109, 113–116, 162, 168–171, 173–175, 181–182
solvent, 31–32, 34, 92–95, 97*f*, 98, 101, 133, 143, 146, 168–170, 173–175, 177–182
synthesis, 34
Organophilic, 1, 52, 59, 63–64, 72–73, 75, 104, 183
Oscillation, 51–52, 68–69, 69*f*
Ostwald ripening, 16
Oxidant
oxidation, 39, 72, 76, 96–98, 100–102, 105–107, 111, 137–139, 143–144, 151, 161–162, 177–178, 181–182
oxidative, 100–102, 163–164, 183
formation, 100–101
polymerization, 96–101, 100*f*, 143, 151
oxidize, 100–102
oxidized, 28, 97, 105, 148*f*
oxidizer, 100–102, 162
oxidizing
agent, 101–102, 104, 165
Oxygen
reduction, 72–73, 75–77, 76*f*, 78*f*
reactions (ORR), 76–77

P

Palladium
chloride, 74
nanoparticles, 74, 173–175, 181
Paramagnetic, 105, 112–113
Particle–particle interactions, 67–68
Particle shape, 64
Particle size, 59, 61, 63–64
Peierls instability, 110
Pendant drop method, 32, 62–63, 168–169
Pentane, 16–17
Perfluorinated, 23, 34, 41–42, 54*f*, 89–90, 132–134, 143–144, 145*f*, 155–156
Permeability, 10, 76–77, 166
Permittivity, 32–33, 64
Peroxide, 177–178
Petroleum, 1–2, 16–17, 30, 168–170, 175–176
pH, 49–50, 53–54, 67, 107–108, 179
Pharmaceutical, 30
Phase diagram, 112–113, 113*f*
Phase separation, 41–42, 56–57
Phase transfer, 23, 28–29, 182–183
catalyst, 9, 113–114, 177, 181–182
Phase transition, 35, 60, 166
Phosphate, 27*f*, 41–42, 64–65
Phospholipid, 6–7, 14–15, 54–55, 77, 78*f*, 98
Phosphorescence, 132, 134, 135*f*, 137
Photo-current, 164
Photoexcited state, 8*f*
Photo-functional, 90
Photoinduced depolarisation, 7
Photo-induced hole, 164
Photoinhibition, 7
Photoluminescence, 133–141, 143–149

Photophysical properties, 61, 91, 122–123, 153–154
Photopolymerized, 152–153
Photo-responsive, 164
Photonics, 78, 131–132, 152–153
Photosynthesis, 12–13, 28
Photosynthetic system II, 26–27
Physical chemistry, 49, 96
Physical properties, 2–3, 37–38, 104–110, 117, 122–123, 153
Physical science, 17–18
Physical thickness, 39–40
Physicochemical, 11–12, 54, 133–134
Physics, 91, 111, 136, 183
Pickering, 176, 179–181
  emulsion, 29, 59, 65, 66*f*, 173–176, 174*t*, 175*f*, 178*f*
Pi–pi, 110
Planetary, 1–2, 13–15
Plasma, 4, 6, 13
  frequency, 111–112
Plasmon, 56–57, 65–68, 69*f*
Plasmonic nanoparticles, 67–71
Plastic, 89–95
Plate method, 32
Platinum, 79, 101–102, 142, 161–162
Pluronic
  F127, 52, 53*t*
  F68, 52, 53*t*
Polar
  amine, 96–97
  behavior, 38–39, 155
  group, 7, 99
  head, 14–15, 52
  media, 9–10, 179–181
Polarizable, 72
Polarization, 70–71
Polarity, 2, 9–10, 23, 39, 44–45, 99–100
Polyacetylene, 90–91, 91*f*, 96, 110
Polyacrylic acid, 99
Polyamide (PA), 91–92, 92*f*, 119*f*
Polyamidoamine, 179–181
Polyaniline (PANI), 39, 91, 91*f*, 96–100, 133–134, 141–142, 163
Polyaromatic, 168–169
Polybutadiene (PBD), 61
Polycarbonate (PC), 92–94
Polycation, 100–101
Polycondensation, 12, 92–94, 113–114, 115*f*, 183
Polyelectrolyte, 39
Polyester, 89–92
Polyethylene
  glycol, 7–9, 61–62, 102–103, 172
    ubiquinol sebacate, 172*f*
  oxide, 52
Polyisobutylene, 90*f*
Polyisoprene, 89
Polymer
  addition, 41–42
  capped nanoparticles, 67
  chain, 98, 100–102, 105–110, 164
    growth, 97
  chemistry, 183
  layer, 104
  matrix, 92–95, 101–103
  reaction, 97
Polymerase, 10–11
Polymeric nanomaterials, 45, 78, 91–92, 96–97
Polymerization, 10–11, 24–25, 54–55, 90, 98–101, 114–116, 118, 119*f*, 120*f*, 143, 146, 151, 163–164
Polymethyl methacrylate (PMMA), 41–42, 61, 63, 74, 90*f*
Polyoxometalates (POM), 164, 176–178, 181–182
Polyoxyethylene
  cetyl ether–water, 79
  cetyl ether–hexadecane–water, 79
Polypropylene (PP), 89, 90*f*
  oxide, 52
Polypyrrole (PPy)
  black, 100–101
  carbon, 162
  nanostructures, 104–105
Polystyrene (PS), 31–32, 31*f*, 42–44, 61, 89, 139–141
Polytetrafluoroethylene, 89
Polyurethane (PU), 91–92
Polyvinyl acetate, 90*f*, 167
Polyvinyl chloride, 90*f*
Polyvinyl ethyl ether, 167
Poor-in-colloid phase, 41–42
Poor-in-polymer phase, 41–42

Porous
nanostructure, 76
silicone, 76
structure, 77, 106–107
porosity, 107–108, 114–116, 119, 154, 165
Porphyrin, 7, 75–77, 120*f*
Positive charge, 110
Positive response, 104–105
Potassium, 55–56, 142, 179
Potential, 7, 13–14, 25–27, 29–30, 39–40, 49–50, 55–56, 64, 70–73, 90, 112–113, 132–133, 142, 163–164, 179
Powder, 59, 121, 153–154
Power, 12, 42–44, 108, 161–162, 164–165
density, 108
Prebiotic chemistry, 1–6
Precipitate
precipitation, 16, 56–57, 64–65
kinetics, 109–110
Precursor, 3–6, 13, 61–62, 91–92, 181
salt, 74
Pressure, 1–3, 14, 15*f*, 16, 18, 29–30, 34–38, 38*t*, 40, 44–45, 64, 131–132, 171, 173–175
coefficient, 39–40
Primordial atmosphere, 4
Production, 12–13, 89–90, 131–132, 137, 144–146, 166, 175*f*
Propagation, 4
Protein, 5–7, 10–11, 49–50, 77, 92–94, 131–132, 179, 180*f*
Protonated, 104–105, 163
Protonation, 39, 96, 111
Proton-coupled electron transfer, 72–73
Purified, 94, 100–101
Pyrolysis, 16, 173–175
Pyrrole
monomer, 100–102, 162
polymerization, 99–101

Q

Quadrupole
molecule, 34, 36*f*
moment, 34, 38–39
Quartz
surface, 37–38
system, 37–38
Quasi–D array, 69
Quaternary, 181–182
ammonium, 172–173, 177–178

R

Radiation, 3–4, 71*f*, 137, 150*f*, 155
Radiative, 143–144
Radiolysis, 123
Ralstonia eutropha, 12–13
Raman spectroscopy, 52, 59, 72, 74, 94
Rapid diffusion, 96, 106–107
Rare earth metal, 132–133, 176
Reaction
conditions, 3–4, 121–122, 173–175
rate, 9–10, 26–27, 137–139, 183
vessel, 5–6, 9–10
Reactive chemical environment, 10
Reactive plasma, 4
Recovery, 166, 171, 178
efficiency, 167
Recycling, 181
stability, 181
Red dye, 182
Redox
couple, 72, 73*f*
process, 49, 74
reaction, 26–29, 72–75, 100–101, 161–162
species, 70–71
Reducing agent, 72–73, 102–103, 106–107
Reduction, 28–29, 58*f*, 60–61, 63–64, 66*f*, 74–75, 77, 106–107, 168–169
Reflection
reflectance, 63–64, 70–71, 111–112, 114–116
reflected, 4, 68
Refolding process, 11*f*
Refraction, 53–54, 69–70
Refractive index, 53–54, 69–70
Reproduction, 1, 5–6
Repulsion, 41, 51–52, 60, 63–65, 79
Repulsive, 51–52

Research, 23, 33–37, 69, 71–73, 79, 89–90, 97, 104–106, 161–162, 165–166
Resistance behavior, 112–113
Resistivity, 94–95, 108–109, 111–112
Resonance, 11–12, 35–37, 51–52, 56–57, 68, 69*f*, 105, 143–144, 177
Response
  magnitude, 106–107
  time, 106–107
Reverse osmosis, 92–94
Rheological properties, 60
Rheological sheer thinning, 65
Rheometer, 65
Rhodium, 101–102
  catalysed asymmetric conjugate addition, 102–103
Ribonucleic acid (RNA), 5–6, 5*f*, 10–11
Ring method, 32
Ring
  opening, 182
    polymerization, 61–62
  reaction, 182
Rock, 16, 37–38, 166–167, 171
  permeability, 166
Rod, 32, 165
  method, 32
Room temperature, 7–9, 14, 72–73, 114–116, 121, 143–146, 153–154, 179
Rubber, 89–90

S

Safety, 182
Salinity, 6, 64–65, 131–132, 167, 175–176
Salt, 2, 49–51, 60, 74, 96–98, 102–106, 109, 135*f*, 154, 171, 175*f*, 176, 179
Scanning electrochemical microscopy, 70
Scanning transmission electron microscopy (STEM), 58*f*
Scanning tunnelling microscopy, 70
Scattering, 41–42, 49–50, 68–69, 114–116, 141
  length, 53–54
Secondary growth, 96–99
Self-assembly, 1, 6, 18, 45, 56–57, 57*f*, 60, 67, 76–77, 143, 154, 161, 182–183, 193
Self-assembled, 18, 50, 59–60, 65–71, 78, 155, 161, 163, 182–183
Self-assembling, 1, 60, 153–154
Self-stabilized dispersion polymerization, 111–112
Semiconductive, 111
  semiconductor, 79–80, 111, 155–156, 172
Semicrystalline, 108–109
Sensitivity, 104–107, 139, 141
Sensor, 96, 104–108, 133, 142, 182–183
Separation, 7, 44, 56–57, 58*f*, 66*f*, 98, 102–103, 141, 143–144, 154, 164, 167, 171–172
Shape, 23, 25, 32–34, 39–40, 54, 60–61, 64, 69, 90, 104–105, 143, 161–162
Sheer-thinning behavior, 65
Silica
  nanoparticle, 67, 175–176
  synthesis, 61–62
Silicic acid, 61–62
Silicone, 67, 76
  oxycarbides, 165–166
Silver
  cation, 100–101, 179
  nanocomposite, 101–102, 113–114
  nanoparticle, 52, 52*f*, 56–57, 68–69, 101–103, 164, 183
  nitrate, 100–102
Single-stranded deoxyribonucleic acid, 154–155
Single-walled carbon nanotube, 173–175
Size, 25, 35–37, 41–42, 54, 59–61, 69, 92–94, 102–105, 117–118, 153–154, 161–162, 182
Small-angle neutron scattering (SANS), 49–50, 169
Sodium chloride, 60
Sodium dodecyl sulfate (SDS), 7–9, 12, 44, 49–50, 52, 53*f*, 53*t*
Sodium dodecylbenzenesulfonate, 99
Soft, 70–71, 123, 164–165

Soft (*Continued*)
reaction, 98
Solar energy
conversion, 161
resources, 12
Solar fuel generator, 77
Solar power, 12
Solar radiation, 4
Solid–liquid interface, 25, 32
Solid-ordered, 6
Solid-state
packing, 108–109
physics, 111
reaction, 153–154
reactivity, 153–154
Soluble
solubilization, 171–172
solubilised, 100–103, 143–144
solubility, 25, 34–39, 44–45, 96, 99–101, 121, 131–132, 135*f*, 167, 173–175
Solute mass transfer, 16
Solution, 5*f*, 11*f*, 24–25, 27–28, 60, 64–65, 102–103, 113–114, 163, 169, 179, 180*f*
Solvent, 2, 9–11, 13–14, 15*f*, 24*f*, 33–39, 70, 92–94, 98, 100–101, 104, 110–111, 116–117, 137, 146, 169–170
Sonogashira coupling reaction, 9, 72, 181
Soybean oil, 176, 178
Sp2, 57*f*, 112–113
Space, 11–12, 16, 96–98, 123, 182–183
Spacing, 54–55
Specific capacitance, 108, 163–165
Spectroscopy, 7, 11–12, 26–27, 35–37, 45, 51–52, 59, 72, 74, 94, 143–144, 151, 175–176
Spectroscopic, 39, 51–52, 59, 151
Spherical, 3–4, 33, 54, 61, 63, 68–69, 69*f*, 99, 102–103, 106–107, 118, 133
Stacking, 74
Storage modulus, 60, 103–104
Storage of capacitive charge, 163–164
Strecker synthesis, 6
Strength, 6, 49, 53–54, 60, 92–94, 103–104, 114–116
Stress–strain, 93*f*
Structure, 7, 15, 23, 41–45, 50–51, 53*f*, 55–56, 76–77, 139, 142–144, 146–148, 148*f*, 161, 177–178
Subcritical, 37–38, 166
Substrate, 7–9, 25–26, 67, 113–116, 121, 142, 153–156, 163, 176
Sudan I, 182
Sugar, 38–39, 171–172
Sulfonate, 11–12, 101
Solfonated β-cyclodextrin, 101
Sulfonyl fluoride vinyl ether, 176
Sulfur, 168, 172, 177–178
dioxide, 11–12
Sulfuric acid, 171–172, 182
Sun, 4
Supercapacitor, 108, 161–165
Supercritical
carbon dioxide, 34–39, 44–45, 165–166
drying, 34–35, 165
fluid, 23, 25–26, 34–39, 45, 165–172
state, 34–35, 166
Superhydrophobic coating, 34, 90
Surface
area, 10–11, 24–25, 106–108, 161–165
chemistry, 173–175
concentration, 168–169
coverage, 70, 71*f*, 170–171
energy, 67
enhanced Raman scattering (SERS), 52, 68
plasmon, 56–57, 65–68, 69*f*
pressure, 64
tension, 23, 29–32, 34–39, 40*f*, 53–54, 57, 61, 67, 175–176
Surfactant, 3–4, 7–9, 30*f*, 44, 49–52, 61–62, 92–94, 98–99, 118, 146–148, 165, 174*t*, 178
Swelling, 106–107
Synthesis, 1, 6, 13, 23, 31*f*, 56–57, 67, 72, 76, 79, 89–90, 92–94, 96, 98, 100*f*, 101–102, 107–108, 116–118, 122–123, 141–142, 153–154, 161, 164, 182–183, 193
Synthesized, 5–11, 13, 17–18, 23–25, 64, 68–69, 72, 89, 91–94, 99, 108, 111–112, 118, 131–132, 152–153, 155–156, 161, 164–165, 177–178, 181

T

Teflon, 25–26, 89–90, 137–139
Temperature
composition, 2–3
range, 111–112
Template, 76, 78–79, 98–99, 102–103, 123, 151, 164–165, 173–175
templating, 165
Tensile modulus, 92–94
Tension, 7–9, 23, 29–34, 37–45, 38*t*, 51–52, 60–67, 95, 99–100, 118, 143, 144*f*, 155–156, 167–170
Tergitol NP-8, 172
Terpolymer, 61–63, 79
Tetrahydrofuran, 62–63, 171–172
Tetramethylammonium tetraphenylborate, 78*f*
Tetramethylene *bis*(chloroformate), 91–92
Tetramethylethylenediamine, 7–9
Tetraoctylammonium bromide, 181
palladium nanoparticle, 72
Thermal
energy, 64–65
fluctuation, 41–42, 65
induced fluctuations, 39–40
recovery, 166
stability, 161–162
treatment, 121–122, 131–132, 153–154
Thermodynaminc, 35–37, 40, 54, 109–110
Thermogravimetric analysis, 101–102, 175–176
Thermoplastic
polymers, 89
Thermosensitive, 175*f*
Thickness, 23, 39–45, 49–50, 54–55, 69, 71, 106–107, 114–119, 121–123, 154, 183
Thin film, 28, 45, 49, 60–71, 90, 92–94, 92*f*, 100–101, 106–107, 110–111, 113–114, 116–117, 122–123, 141–143, 155–156, 183
Three-dimensional
arrays, 168–169
network, 101
Three-phase, 51–52, 59, 178
Time
resolved fluorescence, 7, 12, 148*f*, 149, 150*f*
response, 106–107
Titan, 14–15, 15*f*
Titania, 176
Titanium dioxide, 181
Toluene
water emulsion, 72
water interface, 63–64, 139–141
Toughness, 92–94, 103–104
Toxic, 34, 141
Transesterification, 176
Transfer, 23–25, 28–29, 59, 74, 77, 136, 155–156, 179, 181–182
Transformation, 16, 31–32, 31*f*, 64
Transition
metal, 11–12, 133, 152, 161–162, 176
temperature, 6, 54–55
Transmission electron microscopy (TEM), 101–102, 155
Transmittance, 63–64, 71
Trapping, 42–44, 181–182
energy, 64
Triblock copolymer, 52, 61–62, 78–79, 176
Triblock terpolymer, 79
Triethylamine, 133, 179–181
Triple-phase boundaries, 51–52
Triton X-100, 12, 52, 53*f*, 53*t*, 102–103, 146–148, 148*f*
Tunable, 35, 50, 90, 155–156
thickness, 183
Tungsten, 177–178
sulfide, 152–153
Tuning, 90, 105, 179
Tween 20, 92–94
Two dimensional
array, 64, 163
entity, 39–40
restricted space, 182–183
space, 96–97, 99, 123, 152
Two-photon absorbance, 61

U

Ultrasonic, 56–57
Ultraviolet
  irradiation, 155
  radiation, 4
  -visible
    spectra, 99, 149
    spectrum, 110–111
Underground oil, 17–18, 34–35, 166–167
Underground rock formation, 166–167
Underground water, 64–65, 131–132, 167
Urey–Miller, 5–6, 5*f*, 13

V

Vacuum, 57
Van Der Waals, 35, 64–65, 154
Vanillin alcohol, 173–175
Vaporization, 2, 13
Vapor
  pressure, 44–45
  solution interface, 60
Vertical structure, 169*f*
Vesicle, 10–12, 11*f*, 15*f*, 50–51, 51*f*, 54–57
Viscous
  gel, 31–32
  viscosity, 24–25, 35, 44–45, 166–167, 171
Volume, 2, 29–30, 32–34, 74, 106–107
  fraction, 33, 79
Voyager 1, 15*f*
Vulcan XC-72, 161–162
Vulcanization, 89

W

Water
  affinity, 104
  -benzene, 92*f*
  -chloroform, 43*f*, 77, 92–94, 143–144, 177–178
  -chloroform-cyclohexane system, 95
  content, 6
  -cyclohexane, 39, 62–63
  -dichloromethane, 63–64, 76, 144*f*
  -fluorocarbon, 34, 147*f*
  -heptane interfacial system, 25–26
  -hydrocarbon interface, 31, 34
  -*n*-decane, 39
  -*n*-heptane, 39
  -*o*-dichlorobenzene, 144*f*
  -oil, 28, 30, 31*f*, 34, 50–51, 51*f*, 59, 62–64, 71, 78, 119*f*, 139–141, 167–171, 169*f*, 179
  -organic synthesis, 34
  -perfluoroether, 144*f*
  phase, 43*f*, 64, 98, 114–116, 171–172
  receding angle, 37–38
  -silicone, 29–30
  -soluble, 44–45, 68–69, 177–178, 177*f*
  -toluene, 61, 63–64, 72, 139–141, 140*f*, 148–149
  -1,2 dichloroethane, 65–67, 77
Waveguide properties, 165
Weight, 7–9, 91–92, 96, 98, 104, 169, 176, 179
Wettability, 37–38, 116–117, 166, 173–176
Wettable ring, 32
Wetted, 37–38
Wide-angle X-ray scattering, 54–55

X

X-ray diffraction (XRD), 92–94, 108–109, 111
X-ray reflectivity, 49–50, 60
X-ray scattering, 41–42, 60, 169

Y

Yield, 5–6, 5*f*, 9, 16–17, 72, 92–94, 96, 98, 103–104, 114, 121, 132–139, 141–142, 155, 172–175, 181
Young's modulus, 67–68, 103–104

Z

Zero bandgap semiconductor, 112–113
Zinc, 7–9
Zwitterionic, 10–11

Kolibri 360 GmbH
Ferdinand-Jühlke-Straße 7
99095 Erfurt, Deutschland
produktsicherheit@kolibri360.de